To Bri

With be

Eric H...

Monographs on Endocrinology

Volume 7

Edited by

F. Gross, Heidelberg · A. Labhart, Zürich
T. Mann, Cambridge · L. T. Samuels, Salt Lake City
J. Zander, München

E. W. Horton

Prostaglandins

With 97 Figures

Springer-Verlag Berlin · Heidelberg · New York 1972

Professor Eric William Horton, D. Sc., F.R.C.P.E.

Department of Pharmacology, University of Edinburgh
1 George Square, Edinburgh EH8 9JZ/Great Britain

ISBN 3-540-05571-1 Springer-Verlag Berlin Heidelberg New York
ISBN 0-387-05571-1 Springer-Verlag New York Heidelberg Berlin

Preface

This book was written at the invitation of Dr. H. Götze of Springer-Verlag for the series "Monographs in Endocrinology". It is not a comprehensive account of the prostaglandins but has been written with a deliberate emphasis upon those aspects of the field in which I am particularly interested and to which, in some cases, I have made a contribution.

I am grateful to Miss E. Pfisterer and her colleagues of Springer-Verlag for their excellent work. I should also like to thank my wife without whose patience, encouragement and help this book would never have been completed.

Finally this is an appropriate time to express my sincere gratitude to those scientists who over the years have given me samples of prostaglandins — namely Professors S. Bergström and B. Samuelsson of the Karolinska Institute, Stockholm, Professor D. A. van Dorp of the Unilever Research Laboratories, Vlaardingen and Dr. J. E. Pike of the Upjohn Company, Kalamazoo. Without their help work in this field would have been extremely difficult.

E. W. H.

Contents

I. Introduction

1. Discovery of Prostaglandin

The credit for the discovery of the prostaglandins rightly belongs to the Swedish scientist, U. S. VON EULER. It is true that other workers had observed pharmacological effects with semen and prostatic extracts which can now, with hindsight, be attributed to the presence of prostaglandins (for example BATTEZ and BOULET, 1913; KURZROK and LIEB, 1930; COCKRILL, MILLER and KURZROK, 1935; GOLDBLATT, 1933, 1935), but it was VON EULER (1934, 1935 a, 1936, 1939) who established beyond doubt that the active principle, which he named prostaglandin, belongs to a completely new group of naturally occurring substances; furthermore, it was at VON EULER's suggestion that Professor SUNE BERGSTRÖM in 1947 took up the problem of prostaglandin purification. This led to the isolation of the first two prostaglandins in 1960 and so a vast new field of chemical, biological and clinical importance was opened up.

VON EULER (1934) observed that human semen and extracts of sheep vesicular glands lower arterial blood pressure on intravenous injection and stimulate various isolated intestinal and uterine smooth muscle preparations. He showed that the active principle, prostaglandin, was a lipid soluble acid and thus differed chemically from all other known subtances with similar biological activity, for example, histamine, acetylcholine and adenylic compounds (VON EULER, 1935 a, 1936). GOLDBLATT (1935) independently distinguished the active principle also from substance P. VON EULER made use of prostaglandin's physico-chemical properties in preparing an extract for further biological work. He extracted sheep vesicular glands with ethanol and after evaporation to dryness, partitioned the residue between ether and water at acid and alkaline pH. This partially purified prostaglandin (PG) was stable for many years.

2. Pharmacological Properties

PG lowered the systemic arterial blood pressure of the urethanised rabbit on intravenous injection (VON EULER, 1935 a). It had little or no action on the isolated perfused heart of the rabbit, but increased flow through the perfused hind limb and kidney and decreased flow through the pulmonary vascular bed (VON EULER, 1939). It thus seemed likely that the depressor action of PG in the rabbit could be attributed to changes in vascular resistance rather than to an effect upon the heart. Pressor responses to adrenaline were also reduced by PG.

VON EULER (1939) observed that the depressor response to PG was more rapid in onset after an injection into the femoral vein than after an injection into the femoral artery, suggesting the possibility of a site of action in addition to a direct vasodilator effect on the hind limb. Injection into the cat portal vein increased portal venous pressure and decreased systemic arterial pressure. There was also decreased flow on injection of PG into the perfused cat kidney. On the cat heart-lung preparation PG had little action (VON EULER, 1939).

Other cardiovascular effects of PG observed at that time were its constrictor action on the perfused human placental vessels (VON EULER, 1938) and its positive inotropic and chronotropic effects on the isolated frog heart.

Prostrate, seminal vesicles and seminal fluid of the Rhesus monkey contain a depressor substance which lacks the smooth muscle stimulating activity of PG. VON EULER (1935 b) named the active principle, vesiglandin. Tissues and organs of several species were tested for PG-like activity (VON EULER and HAMMARSTRÖM, 1937). Some activity was found in the ovary. However, in no organ could PG be detected in a concentration approaching that found in human seminal plasma.

3. Prostaglandin and Reproduction

ELIASSON and his colleagues working in VON EULER's laboratory made several important contributions to the role of PG in reproductive physiology. ELIASSON (1959) confirmed that PG, like human semen, inhibits the spontaneous contractions of isolated myometrial strips from the non-pregnant woman and showed that the preparation is most sensitive at the time of ovulation (BYGDEMAN and ELIASSON, 1963). ELIASSON and POSSE (1960) found that PG administered intravaginally stimulates contractions of the non-pregnant human uterus at the time of ovulation. ELIASSON (1957) established by making comparisons on various biological preparations that the PG from human seminal fluid was similar to that in the sheep vesicular gland. He also observed that when minced sheep vesicular glands were incubated in phosphate buffer at 37°, their PG content increased 10-fold in a few minutes. Furthermore the amount formed could be increased by the addition of phospholipase A. Incubation of human seminal fluid with or without enzyme did not increase the yields of prostaglandins (ELIASSON, 1959).

Finally, ELIASSON (1959) made the important observation, by fractionation of ejaculates, that human seminal PG is secreted mainly by the seminal vesicles not by the prostate. Thus the original assumption which gave rise to the name, prostaglandin, was proved incorrect. By that time however, pure prostaglandins were being isolated and the name soon became established in the literature.

4. Isolation and Structure

The impetus for further biological work resulted from the elegant isolation and chemical characterisation achieved by BERGSTRÖM, SJÖVALL, SAMUELSSON and their co-workers at the Karolinska Institute. BERGSTRÖM and SJÖVALL (1960 a, b) isolated two compounds which behaved differently on partition between ether and an aqueous phosphate buffer. The one more soluble in ether was called prostaglandin E (PGE), the other more soluble in phosphate buffer (in Swedish spelt with an "F") was called prostaglandin F (PGF). These compounds were assigned the empirical formulae $C_{20}H_{34}O_5$ and $C_{20}H_{36}O_5$ respectively.

It was soon discovered that unsaturated analogues of these prostaglandins are present in human semen and in other tissues, notably the lungs.

In 1962 and 1963 BERGSTRÖM and his co-workers announced the chemical structure of several naturally occurring prostaglandins (BERGSTRÖM, RYHAGE, SAMUELSSON and SJÖVALL, 1963). Since then research in this field has advanced rapidly.

5. Prostaglandin-like Substances

In the meantime, other workers had discovered smooth muscle stimulating lipids in numerous tissues. Frog intestine spontaneously releases darmstoff (VOGT, 1949). Rabbit iris contains irin (AMBACHE, 1957, 1959) which may be the chemical mediator responsible for the atropine-resistant miosis resulting from antidromic stimulation of the trigeminal nerve. The menstrual stimulants are a group of lipids extracted from human menstrual fluid (PICKLES, 1957). Several workers have reported the presence of smooth muscle-stimulating lipids in the brain (AMBACHE and REYNOLDS, 1960, 1961; KIRSCHNER and VOGT, 1961; TOH, 1963). LEE, COVINO, TAKMAN and SMITH (1965) published a paper on the depressor lipid in rabbit kidney medulla, which they called medullin. All these substances are lipid soluble, all behave like organic acids and all have pharmacological actions on smooth muscle.

Recent work suggests that the presence of one or more prostaglandins can account for much if not all the biological activity of frog darmstoff (VOGT, SUZUKI and BABILLI, 1966), human menstrual fluid (EGLINTON, RAPHAEL, SMITH, HALL and PICKLES, 1963), sheep iris extracts (ÄNGGÅRD and SAMUELSSON, 1963), ox brain lipids (SAMUELSSON, 1964) and medullin (LEE, CROWSHAW, TAKMAN, ATTREP and GOUGOUTAS, 1967).

6. Nomenclature

The basic 20-carbon skeleton of the prostaglandins has been named prostanoic acid (Fig. 1). The correct chemical name of all prostaglandins, their metabolites and analogues can be derived by refer-

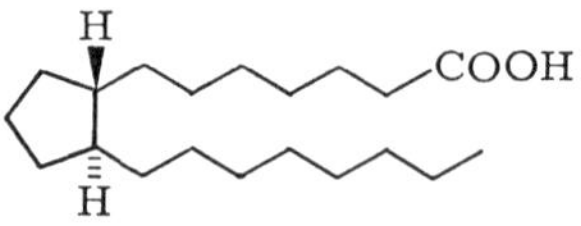

Fig. 1. Prostanoic acid

ence to this structural formula. These chemical names although precise are long and tedious to use; for the major prostaglandins trivial names have been retained.

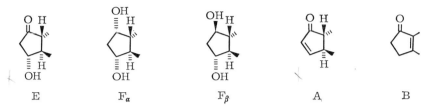

E F_α F_β A   B

Fig. 2. Structural differences between prostaglandins of the E, F, A and B
series

PGE$_1$

PGF$_{1\alpha}$

PGE$_2$

PGF$_{2\alpha}$

PGE$_3$

PGF$_{3\alpha}$

PGA$_1$

PGB$_1$

PGA$_2$

PGB$_2$

Fig. 3

Four series of natural prostaglandins have so far been described, designated by the letters E, F, A and B corresponding to differences in the five membered ring (Fig. 2). All the "primary" prostaglandins (Fig. 3) are hydroxylated in the 15 position and contain a 13, 14-*trans* double bond. The degree of unsaturation of the side chains is indicated by the subscript numeral after the letter, thus prostaglandins E_1, F_1,

Table 1. *Names and designations of some prostaglandins*
(see Fig. 3 for structural formulae)

Trivial name	Abbreviation (used in this book)	Chemical name	Andersen's designation
Prostaglandin E_1	PGE_1	$11\alpha,15\alpha$-dihydroxy-9-oxo-13-*trans*-prostenoic acid	PG $(E\alpha\alpha)_1$
Prostaglandin E_2	PGE_2	$11\alpha,15\alpha$-dihydroxy-9-oxo-5-*cis*-13-*trans*-prostadienoic acid	PG $(E\alpha\alpha)_2$
Prostaglandin $F_{1\alpha}$	$PGF_{1\alpha}$	$9\alpha,11\alpha,15\alpha$-trihydroxy-13-*trans*-prostenoic acid	PG $(\alpha\alpha\alpha)_1$
Prostaglandin $F_{1\beta}$	$PGF_{1\beta}$	$9\beta,11\alpha,15\alpha$-trihydroxy-13-*trans*-prostenoic acid	PG $(\beta\alpha\alpha)_1$
Prostaglandin $F_{2\alpha}$	$PGF_{2\alpha}$	$9\alpha,11\alpha,15\alpha$-trihydroxy-5-*cis*-13-*trans*-prostadienoic acid	PG $(\alpha\alpha\alpha)_2$
Prostaglandin A_1	PGA_1	15α-hydroxy-9-oxo-10,13-*trans*-prostadienoic acid	PG $(A\varDelta\alpha)_1$
Prostaglandin B_1	PGB_1	15α-hydroxy-9-oxo-8 (12),13-*trans*-prostadienoic acid	PG $(B-\alpha)_1$

Footnote On Andersen's designation. Configurations at C-9, C-11 and C-15 are indicated, in that order, by α or β within parentheses. Subscripts indicate the degree of unsaturation in the corresponding E-type prostaglandin. When no configuration assignment is required, the designations are: K = oxo, $\varDelta$ = double bond, and — (dash) for no substituent. In 9-oxo compounds E, A or B replaces K. Prostaglandins with *cis*-orientated side chains are designated by prefixing iso-to the abbreviation used for the C-8 epimer. The antipodes and racemates are designated by the prefixes ent- and rac- respectively.

A_1 and B_1 have only the *trans* double bond, prostaglandins E_2, F_2, A_2 and B_2 have in addition, a *cis* double bond in the 5, 6 position while prostaglandins E_3, F_3 and B_3 have an additional *cis* double bond in the 17, 18 position. Chemical reduction of a prostaglandin E yields two isomeric alcohols F α and F β.

In this monograph prostaglandins will be referred to where possible by an abbreviation e. g., PGE_1, PGF_{2a}, PGB_3, etc. (see Table 1). Where ambiguity might occur the full chemical name is used.

An alternative system of abbreviations which takes into account stereochemical configurations has been introduced by ANDERSEN (1969). This new designation is useful when discussing the effects of stereochemical changes upon pharmacological activity or chromatographic behaviour. The essential features are described in the footnote to Table 1, but the system will not be used in this monograph.

7. Prostaglandin Literature

Well over a thousand papers have now been published on various aspects of prostaglandin research. Many are not referred to in this book. For those readers who wish to keep abreast of this rapidly expanding literature, the prostaglandin bibliography prepared and published by the Upjohn Company, Kalamazoo, is invaluable. Recent reviews have been written by BERGSTRÖM (1967), BERGSTRÖM, CARLSON and WEEKS (1968), VON EULER and ELIASSON (1967), HINMAN (1967), PICKLES (1967, 1969), VON EULER (1968), HORTON (1968, 1969), RAMWELL and SHAW (1970) and RAMWELL, SHAW, CLARKE, GROSTIC, KAISER and PIKE (1968).

Earlier literature has been reviewed by VOGT (1958), ELIASSON (1959) and VON EULER (1966). The proceedings of several symposia have been published (PICKLES and FITZPATRICK, 1966; BERGSTRÖM and SAMUELSSON, 1967; RAMWELL and SHAW, 1968, 1971; MANTEGAZZA and HORTON, 1969).

References

AMBACHE, N.: Properties of irin, a physiological constituent of the rabbit's iris. J. Physiol. **135**, 114—132 (1957).
— Further studies on the preparation, purification and nature of irin. J. Physiol. **146**, 255—294 (1959).

AMBACHE, N., REYNOLDS, M.: Ether-soluble active lipid in rabbit brain extracts. J. Physiol. **154**, 40 P (1960).

— — Further purification of an active lipid from rabbit brain. J. Physiol. **159**, 63—64 P (1961).

ANDERSEN, N. H.: Preparative thin-layer and column chromatography of prostaglandins. J. Lipid Res. **10**, 316—319 (1969).

ANGGÅRD, E., SAMUELSSON, B.: Smooth muscle stimulating lipids in sheep iris. The identification of prostaglandin $F_{2\alpha}$. Biochem. Pharmac. **13**, 281—283 (1964).

BATTEZ, G., BOULET, L.: Action de l'extrait de prostate humaine sur la vessie et sur la pression artérielle. C. R. Séanc. Soc. Biol. **74**, 8—9 (1913).

BERGSTRÖM, S.: Prostaglandins: Members of a new hormonal system. Science, N. Y. **157**, 382—391 (1967).

— CARLSON, L. A., WEEKS, J. R.: The prostaglandins: A family of biologically active lipids. Pharmac. Rev. **20**, 1—48 (1968).

— SAMUELSSON, B. (editors): Second Nobel Symposium: Prostaglandins. Stockholm: Almqvist and Wiksell 1967.

— RYHAGE, R., SAMUELSSON, B., SJÖVALL, J.: The structures of prostaglandin E_1, $F_{1\alpha}$ and $F_{1\beta}$. J. biol. Chem. **238**, 3555—3564 (1963).

— SJÖVALL, J.: The isolation of prostaglandin F from sheep prostate glands. Acta chem. scand. **14**, 1693—1700 (1960 a).

— — The isolation of prostaglandin E from sheep prostate glands. Acta chem. scand. **14**, 1701—1705 (1960 b).

BYGDEMAN, M., ELIASSON, R.: The effect of prostaglandin from human seminal fluid on the motility of the non-pregnant human uterus *in vitro*. Acta physiol. scand. **59**, 43—51 (1963).

COCKRILL, J. R., MILLER, E. G., KURZROK, R.: The substance in human seminal fluid affecting uterine muscle. Am. J. Physiol. **112**, 557—580 (1935).

EGLINTON, G., RAPHAEL, R. A., SMITH, G. N., HALL, W. J., PICKLES, V. R.: The isolation and identification of two smooth muscle stimulants from menstrual fluid. Nature (Lond.) **200**, 960, 993—995 (1963).

ELIASSON, R.: A comparative study of prostaglandin from human seminal fluid and from prostate gland of sheep. Acta physiol. scand. **39**, 141—146 (1957).

— Studies on prostaglandin. Occurrence, formation and biological actions. Acta physiol. scand. **46**, Suppl. 158, 1—73 (1959).

— POSSE, N.: The effect of prostaglandin on the non-pregnant human uterus *in vivo*. Acta obstet. gynec. scand. **39**, 112—126 (1960).

GOLDBLATT, M. W.: A depressor substance in seminal fluid. J. Soc. chem. Ind., (Lond.) **52**, 1056—1057 (1933).

— Properties of human seminal plasma. J. Physiol. **84**, 208—218 (1935).

HINMAN, J. W.: The prostaglandins. Bioscience **17**, 779—785 (1967).

HORTON, E. W.: The prostaglandins. In: Recent Advances in Pharmacology, 4th ed. Eds.: R. S. STACEY and J. M. ROBSON. London: Churchill 1968, pp. 185—212.

HORTON, E. W.: Hypotheses on physiological roles of prostaglandins. Physiol. Rev. **49**, 122—161 (1969).

KIRSCHNER, H., Vogt, W.: Pharmakologisch wirksame lipoidlösliche Säuren in Hirnextrakten: Isolierung von Lysophosphatidsäure und Gangliosid. Biochem. Pharmacol. **8**, 224—234 (1961).

KURZROK, R., LIEB, C.: Biochemical studies of human semen. The action of semen on the human uterus. Proc. Soc. exp. Biol., N. Y. **28**, 268—272 (1930).

LEE, J. B., COVINO, B. G., TAKMAN, B. H., SMITH, E. R.: Renomedullary vasodepressor substance, medullin: Isolation, chemical characterization and physiological properties. Circulation Res. **17**, 57—77 (1965).

— CROWSHAW, K., TAKMAN, B. H., ATTREP, K. A., GOUGOUTAS, J. Z.: The identification of prostaglandins E$_2$, F$_{2\alpha}$ and A$_2$ from rabbit kidney medulla. Biochem. J. **105**, 1251—1260 (1967).

MANTEGAZZA, P., HORTON, E. W. (editors): Prostaglandins, Peptides and Amines. London: Academic Press 1969.

PICKLES, V. R.: A plain muscle stimulant in the menstruum. Nature (Lond.) **180**, 1198—1199 (1957).

— The prostaglandins. Biol. Rev. **42**, 614—652 (1967).

— Prostaglandins. Nature, Lond. **224**, 221—225 (1969).

— FITZPATRICK, R. J. (editors): Endogenous substances affecting the myometrium. Mem. Soc. Endocr. **14**, 3—144 (1966).

RAMWELL, P. W., SHAW, J. E. (editors): Prostaglandin Symposium of the Worcester Foundation for Experimental Biology. New York: Wiley 1968.

— — Biological significance of the prostaglandins. Rec. Progr. Hormone Res. **26**, 139—187 (1970).

— — (editors): The prostaglandins. Ann. N. Y. Acad. Sci. **180**, (1971).

— — CLARKE, G. B., GROSTIC, M. F., KAISER, D. G., PIKE, J. E.: Prostaglandins. In: Progress in the Chemistry of Fats and Other Lipids, Vol. 9. Ed. R. T. HOLMAN. Oxford: Pergamon Press 1968, pp. 231—273.

SAMUELSSON, B.: Identification of a smooth muscle stimulating factor in bovine brain. Biochim. biophys. Acta **84**, 218—219 (1964).

TOH, C. C.: Biologically active substances in brain extracts. J. Physiol. **165**, 47—61 (1963).

VOGT, W.: Über die stoffliche Grundlage der Darmbewegungen und das Vagusriezes am Darm. Arch. exp. Path. Pharmak. **206**, 1—11 (1949).

— Naturally occurring lipid-soluble acids of pharmacological interest. Pharmac. Rev. **10**, 407—435 (1958).

— SUZUKI, T., BABILLI, S.: Prostaglandins in SRS-C and in darmstoff preparation from frog intestinal dialysates. Mem. Soc. Endocr. **14**, 137—142 (1966).

VON EULER, U. S.: Zur Kenntnis der pharmakologischen Wirkungen von Nativsekreten und Extrakten männlicher accessorischer Geschlechtsdrüsen. Arch. exp. Path. Pharmak. **175**, 78—84 (1934).

VON EULER, U. S.: Über die spezifische blutdrucksenkende Substanz des menschlichen Prostata- und Samenblasensekretes. Klin. Wschr. 14, 1182—1183 (1935 a).

— A depressor substance in the vesicular gland. J. Physiol. 84, 21 P (1935 b).

— On the specific vasodilating and plain muscle stimulating substances from accessory genital glands in man and certain animals (prostaglandin and vesiglandin). J. Physiol. 88, 213—234 (1936).

— HAMMARSTRÖM, S.: Über das Vorkommen des prostaglandins in Tierorganen. Skand. Arch. Physiol. 77, 96—99 (1937).

— Action of adrenaline, acetycholine and other substances on nerve-free vessels (human placenta). J. Physiol. 93, 129—143 (1938).

— Weitere Untersuchungen über Prostaglandin, die physiologisch aktive Substanz gewisser Genitaldrüsen. Skand. Arch. Physiol. 81, 65—80 (1939).

— Introductory survey: Prostaglandin. Mem. Soc. Endocr. 14, 3—18 (1966).

— Prostaglandins. Clin. Pharmac. Ther. 9, 228—239 (1968).

— ELIASSON, R.: Prostaglandins. Medicinal Chemistry Monographs, Vol. 8. New York-London: Academic Press 1967.

II. Extraction, Separation, Identification and Estimation

Prostaglandins have been isolated from various human and animal tissues and fully characterised by rigorous chemical methods. From the data in Table 1 it is evident that these compounds are widely distributed in the body.

The concentrations are generally small, so that extensive chemical analysis may not be possible if only limited amounts of tissue are available. Such minute amounts may be 'identified' by comparing the

Table 1. *Tissues from which prostaglandins have been isolated and identified by full structural elucidation*

	E_1	E_2	E_3	F_{1a}	F_{2a}	F_{3a}	Reference
Seminal plasma (human)	+	+	+	+	+		1, 2
Seminal plasma (sheep)	+	+	+	+	+		3, 4
Vesicular gland (sheep)	+	+	+	+			5, 6, 7
Menstrual fluid (human)		+			+		8
Lung (sheep)		+			+		9, 10
Lung (ox)					+	+	11
Lung (pig, guinea-pig, monkey, human)					+		9, 10
Iris (sheep)					+		12
Brain (ox)					+		13
Thymus (calf)	+						14
Renal medulla (rabbit)		+			+		15, 16

1 BERGSTRÖM and SAMUELSSON (1962)
2 SAMUELSSON (1963)
3 BERGSTRÖM, KRABISCH and SJÖVALL (1960)
4 BYGDEMAN and HOLMBERG (1966)
5 BERGSTRÖM et al. (1962)
6 BERGSTRÖM and SJÖVALL (1960 a)
7 BERGSTRÖM and SJÖVALL (1960 b)
8 EGLINTON et al. (1963)

9 ÄNGGÅRD (1965)
10 BERGSTRÖM et al. (1962)
11 SAMUELSSON (1964 b)
12 ÄNGGÅRD and SAMUELSSON (1964)
13 SAMUELSSON (1964 a)
14 BERGSTRÖM and SAMUELSSON (1963)
15 LEE et al. (1967)
16 DANIELS et al. (1967)

unknown with authentic samples of pure prostaglandins on chromatographic systems and by parallel biological assay. The serious limitations of such methods must be realised. Even if chromatography on several systems and parallel assay on several different tissues suggest that the unknown compound is identical to a particular prostaglandin, the identification must nevertheless still be regarded as tentative. Only negative evidence can be conclusive. For example it may be possible to claim with justification that the unknown is *not* a particular prostaglandin (see HORTON and MAIN, 1967, for a fuller discussion of this problem). Some authors appear to base their identification upon the results from one chromatographic system and a single bioassay — conclusions drawn from such work must be examined critically.

Prostaglandins are also released from numerous sites in response to a variety of stimuli. Again the amounts available for analysis may be very small. This raises many problems. An efficient method of extraction (and a means of estimating its efficiency) is needed; the different prostaglandins must be separated from each other before they can be estimated; a conclusive method of identification applicable to the nanogram scale is essential; and finally there must be a reliable and sensitive method of estimation. The methods currently available for achieving these objectives are described and discussed in this chapter.

1. Solvent Extraction and Partition Procedures

a) Tissues

Prostaglandins can be extracted from a tissue by homogenising it in 4 volumes of 96% ethanol (Fig. 1), centrifuging and extracting the residue again with a further 4 volumes of ethanol (SAMUELSSON, 1963). Other methods of lipid extraction may be used provided that the pH is kept within the range 4 to 8, that the temperature is not raised above 45° and that the procedure is not unduly lengthy. Radioactive prostaglandin should be added to the homogenate to obtain an estimate of recovery by isotope dilution. The possibility that small quantities of prostaglandins may be formed non-enzymatically from their precursor acids during the extraction procedure must be re-

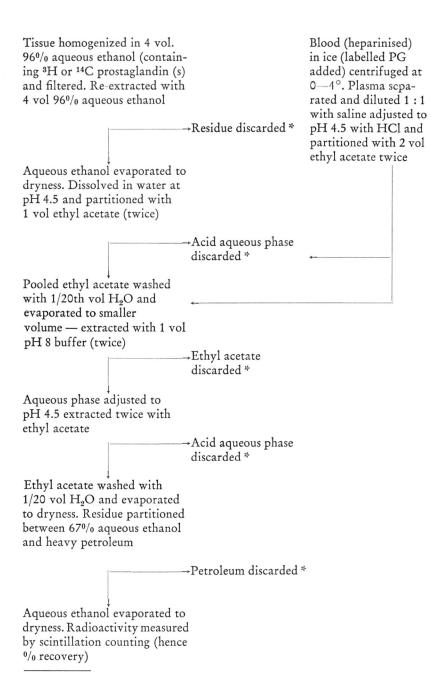

Tissue homogenized in 4 vol.
96% aqueous ethanol (containing ³H or ¹⁴C prostaglandin (s)
and filtered. Re-extracted with
4 vol 96% aqueous ethanol

Blood (heparinised)
in ice (labelled PG
added) centrifuged at
0—4°. Plasma sepa-
rated and diluted 1 : 1
with saline adjusted to
pH 4.5 with HCl and
partitioned with 2 vol
ethyl acetate twice

⟶Residue discarded *

Aqueous ethanol evaporated to
dryness. Dissolved in water at
pH 4.5 and partitioned with
1 vol ethyl acetate (twice)

⟶Acid aqueous phase
discarded *

Pooled ethyl acetate washed
with 1/20th vol H_2O and
evaporated to smaller
volume — extracted with 1 vol
pH 8 buffer (twice)

⟶Ethyl acetate
discarded *

Aqueous phase adjusted to
pH 4.5 extracted twice with
ethyl acetate

⟶Acid aqueous phase
discarded *

Ethyl acetate washed with
1/20 vol H_2O and evaporated
to dryness. Residue partitioned
between 67% aqueous ethanol
and heavy petroleum

⟶Petroleum discarded *

Aqueous ethanol evaporated to
dryness. Radioactivity measured
by scintillation counting (hence
% recovery)

* Measurements of radioactivity may be made on these fractions before
discarding.

Fig. 1. Extraction of prostaglandins E and F from tissues and blood

membered (NUGTEREN, VONKEMAN and VAN DORP, 1967). It may thus be advisable to carry out extractions under nitrogen, particularly if minute quantities of prostaglandins are being extracted from tissues known to be rich in precursor acids.

After evaporation of the alcohol to dryness, the extract is partitioned between water at pH 4.5 (using hydrochloric or citric acid) and an equal volume of ethyl acetate (diethyl ether may be used but it is no better and it has the disadvantages of peroxide formation, high volatility and flammability). The ethyl acetate is separated and the aqueous phase shaken again with a further equal volume of ethyl acetate. The ethyl acetate phases are combined, washed with 1/20th volume of distilled water to remove excess acid and concentrated to a smaller volume using a rotary evaporator (temperature $< 45°$).

The ethyl acetate concentrate is then extracted with an equal volume of pH 8.0 phosphate buffer, the buffer is separated and the ethyl acetate re-extracted with further buffer. The combined aqueous phases are then adjusted to pH 4.5 with hydrochloric (or citric) acid and finally extracted twice with ethyl acetate. The combined ethyl acetate phases are washed with water (1/20 volume) and then evaporated to dryness. The residue in the flask is dissolved by the addition of a small volume of 67% aqueous ethanol and an equal volume of petroleum (the volumes will depend upon the amount and solubility of the residue). After shaking, the ethanol is separated and re-extracted with a further equal volume of petroleum. The aqueous ethanol phase is evaporated to dryness in preparation for chromatography..

b) Blood

Prostaglandins added to blood or plasma are not extracted well by the ethanol method described above (HOLMES, HORTON and STEWART, 1968). It is probable that prostaglandins are absorbed on to and precipitated with the plasma proteins. Better recoveries are obtained by the following method (Fig. 1).

After collection the heparinised blood is cooled quickly in ice, labelled PG is added and the blood is centrifuged at 4°. The plasma is separated, adjusted to pH 4.5 (using a pH meter) by the addition of hydrochloric acid and extracted twice with an equal volume of ethyl acetate. Great care is needed to avoid emulsification. This can be minimised by dilution of the plasma with 1 volume of water or by using 2 to 4 volumes of ethyl acetate for each extraction. If emulsions occur the phases may be separated by centrifugation, but this can be tedious if the volumes are large. The subsequent procedure then follows that described for tissues (above).

Recoveries with this method for PGE_1 and $PGF_{2\alpha}$ (added to blood to give concentrations af 10—100 ng/ml) are between 40 and 55%. With PGA_1 however this method gives very low recoveries (5—10%). Somewhat higher recoveries of PGA_1 are obtained from protein-free Ringer perfusates. For extraction of PGA compounds partition between ethyl acetate and pH 8 buffer should be avoided since these less polar prostaglandins are not extracted efficiently into the aqueous phase. Most of the loss has however occurred before this stage is reached.

Better recoveries for PGE compounds have been reported using a methylal : ethanol (4 : 1) mixture as the first step for the extraction of plasma (HICKLER, 1968). Other pharmacologically active substances are extracted from blood using this procedure and these are troublesome if biological assay is to be used to determine the prostaglandins. This method has not found wide application and is not currently used in this laboratory.

Another method designed to overcome the losses on extraction of prostaglandins from blood is that of UNGER and BENNETT (1971). A slightly modified version (Fig. 2) is used in this laboratory and has given a 78% recovery of ^3H-$PGF_{2\alpha}$ after silicic acid chromatography (initial $PGF_{2\alpha}$ concentration 10 ng/ml blood).

Plasma from heparinised blood is diluted with an equal volume of saline (used to wash the blood cells). The diluted plasma is mixed with an equal volume of ethanol, which results in slight precipitation of plasma proteins. Neutral fat and non-polar fatty acids are removed by partition with petroleum. Formic acid is then added sufficient to make a 1% v/v concentration and the acidified aqueous ethanol extract is partitioned with chloroform. The chloroform layer is evaporated until all traces of formic acid are removed. Further purification by chromatography on silicic acid is of course necessary since pharmacologically active substances other than known prostaglandins are present in the final extract. This method appears to offer many advantages but requires further assessment.

Whether prostaglandins are extracted from tissues, blood or other body fluids, chromatographic purification is essential before any

Heparinised blood (labelled PG added) centrifuged.
Plasma separated. Cells washed with equal vol
saline recentrifuged; plasma and saline mixed with
1 vol ethanol, partitioned with petroleum (twice).

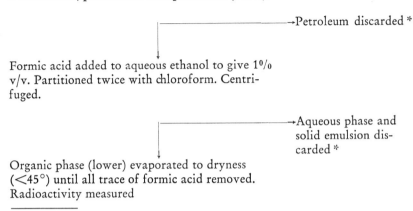

→Petroleum discarded *

Formic acid added to aqueous ethanol to give 1%
v/v. Partitioned twice with chloroform. Centrifuged.

→Aqueous phase and
solid emulsion dis-
carded *

Organic phase (lower) evaporated to dryness
(<45°) until all trace of formic acid removed.
Radioactivity measured

* Radioactivity of these fractions may be measured before discarding.

Fig. 2. Extraction of prostaglandins from blood (Unger et al. method)

quantitative estimations can be made. Gas chromatography of such
crude lipid extracts is unlikely to give results of either qualitative or
quantitative value. Biological assay of the relatively crude lipid
extract obtained by the methods outlined is of doubtful value and
assessment of extraction procedures should not be based upon esti-
mates of the biological activity of such samples. There are several
examples of tissues which contain pharmacologically-active polar
acidic lipids which partition like prostaglandins but which can be
separated from all known prostaglandins by subsequent chromatography.

2. Chromatographic Separation

a) Silicic Acid and Silica Gel Column Chromatography

Chromatography on a column of silicic acid is an effective method
of separating prostaglandins from substances of less and greater
polarity and of separating the prostaglandins themselves into several
broad groups.

Silicic acid is activated by heating for 1 hour at 110°. It is then washed with petroleum and suspended in an ethyl acetate-benzene mixture (1 : 9). The prostaglandin-containing extract is applied in a small volume of the solvent mixture and the column is developed by elution with an ethyl acetate-benzene mixture* of successively increasing ethyl acetate concentration (Fig. 3) or by gradient elution.

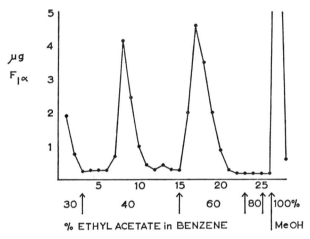

Fig. 3. Chromatography of an extract of chicken brain on a 2 g column of silicic acid. Elution rate 1—1.2 ml/min. 20 ml fractions collected. Ordinate: biological activity assayed on the rabbit jejunum in terms of $PGF_{1\alpha}$. Abscissa: upper line, fraction number; lower line, nature of eluant. (HORTON and MAIN, 1967)

Prostaglandins A and B are eluted first (with the 30% ethyl acetate) then prostaglandins E finally prostaglandins F. Any of these fractions may of course contain prostaglandin metabolites; for example, 19-hydroxy derivatives of the prostaglandins A and B are eluted with the PGF fraction as is 5,7-dihydroxy-11-oxo-tetranor prostadioic acid, the main guinea-pig urinary metabolite of $PGF_{2\alpha}$. Some metabolites and methyl esters of prostaglandins may be eluted with methanol in ethyl acetate.

The properties of silicic acid vary from one batch to another even from the same supplier and so it is essential to check the behaviour of

---

* Care must be taken not to inhale benzene vapour. There may be some advantage in substituting toluene.

authentic prostaglandins by adding labelled prostaglandins as internal standards. If the elution characteristics of an unknown compound on silicic acid chromatography are to be used as part of the evidence of identification, such precautions are particularly important. With some silicic acid the addition of a small percentage of methanol to the ethyl acetate-benzene mixture may be required to elute and separate the prostaglandins.

Silica gel (acid-washed) suspended in chloroform and developed with increasing concentrations of methanol in chloroform is an alternative system to silicic acid. It has the advantage that the solvent system is non-flammable and less toxic. It has been used successfully for gradient elution. Silica gel developed with ethyl acetate-benzene mixtures may be used to separate the PGF epimers formed on $NaBH_4$ reduction of PGE (DANIELS and PIKE, 1968).

b) Sephadex LH-20 Columns

ÄNGGÅRD and BERGKVIST (1970) have described a method for the separation of methyl esters of prostaglandins A (and B), E and F on Sephadex LH-20. The method is used in this laboratory, particularly for further purification of samples immediately prior to gas chromatography. When blood pigments are present in an extract, these are eluted before the PGA fraction — a separation which is difficult to achieve on silicic acid. Unfortunately, this system has not yet been applied to the free acids.

Sephadex LH-20 (Pharmacia) is refluxed with methanol-chloroform (1 : 1), filtered and dried at 40°. It is equilibrated for 3 to

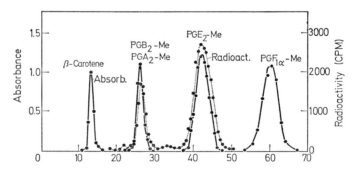

Fig. 4. Chromatographic separation of the methyl esters of PGA_2, PGE_2 and $PGF_{2\alpha}$ on a column of Sephadex LH-20. (ÄNGGÅRD and BERGKVIST, 1970)

4 hours with excess of the eluting solvent mixture, heptane-chloro-
form-cthanol (10-10-1). The gel slurry, after brief evacuation under
suction, is poured into a column (100 cm) and allowed to settle under
free flow. Prostaglandin esters are dissolved in eluting solvent and
applied in 0.2 ml. An elution rate of 2—3 ml/min is appropriate.
Separation of three main groups of prostaglandin esters is shown in
Fig. 4.

c) Amberlyst 15 Resin

One of the most difficult separations to achieve preparatively is
that between prostaglandins differing only in their degree of unsaturation. Thin layer plates of silica gel impregnated with silver
nitrate have been used extensively but these are inapplicable for use
on larger scale.

A sulphonic acid ion exchange resin, Amberlyst 15, impregnated
with silver nitrate has been used successfully to separate PGE_1 (eluted
with ethanol) from PGE_2 (eluted with ethanol containing 5% cyclohexene) (Fig. 5). It is essential to convert all the resin to the silver

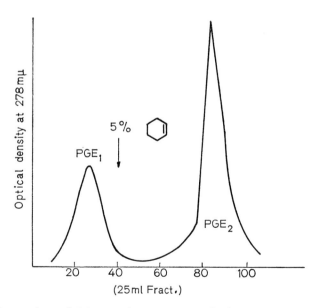

Fig. 5. Separation of PGE_1 and PGE_2 on Amberlyst 15 (Ag^+). Fractions
were assayed by measuring the optical density at 278 nm following treatment with methanolic potassium hydroxide. (DANIELS and PIKE, 1968)

cycle, otherwise acid-catalysed dehydration of PGE to PGA and esterification occur. The same technique has been used for the separation of $PGF_{1\alpha}$ and $PGF_{2\alpha}$ (DANIELS and PIKE, 1968).

d) Reversed Phase Systems

This is probably the most valuable method for the purification of prostaglandins and their metabolites. By selecting the appropriate solvent system, compounds differing either in polarity, chain length or degree of unsaturation can be separated. For example $PGF_{3\alpha}$ can be separated from $PGF_{2\alpha}$ (SAMUELSSON, 1964 b).

Columns of up to 50 g of hydrophobic celite (Hyflo Super Cel) can be conveniently used with a variety of stationary and mobile

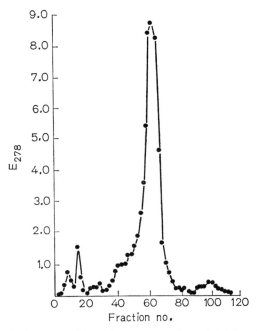

Fig. 6. Reversed-phase partition chromatography of PGA_2. The following conditions were used: column, 4.5 g of hydrophobic Celite, weight of extract containing PGA_2, 17 mg; temperature, 20°; fractions, 2 ml; solvent system, methanol : water : chloroform : 2-ethylhexan-1-ol (9 : 11 : 1 : 1 v/v), upper phase — mobile, lower phase — stationary. Fractions were assayed by measuring the optical density at 278 nm following treatment with N sodium hydroxide. (LEE et al., 1967)

phases. The two phases are thoroughly equilibrated before use. The cclitc must bc carcfully prcparcd by adcquate washing, silanation and drying. A column of 4.5 g supports about 4 ml stationary phase. Mobile phase (50 ml) is added and after thorough mixing the slurry is poured into a column and allowed to settle under gravity or positive pressure. A constant environmental temperature is desirable in order to avoid phase separations. The extract dissolved in mobile phase is loaded in an appropriately small volume (Fig. 6). Details of the method are published by LEE, CROWSHAW, TAKMAN, ATTREP and GOUGOUTAS (1967), NORMAN and SJÖVALL (1958) and HAMBERG (1968).

e) Amberlite XAD-2

This resin can be used for the concentration of prostaglandin metabolites from urine. The procedure is as follows: 75 g of resin forms a column of dimensions 51×2 cm. The column is thoroughly washed with distilled water. Urine (1 litre) is percolated through the column at 5—8 ml/min, 100—200 ml fractions being collected. The column is washed with 100 ml water and then with 4×100 ml of ethanol. Most of the prostaglandin metabolites are eluted in the first 100 ml ethanol and virtually all after 250 ml has passed through. The metabolites may be further purified by solvent partition followed by reversed phase and other forms of chromatography.

f) Thin Layer Chromatography

Numerous solvent systems have been described for the thin layer chromatographic separation of prostaglandins and their metabolites. Some of these are listed in Table 2. Thin layer chromatography has the advantages of simplicity and speed. It is useful for small samples and plates can be scanned for radioactivity (Fig. 7).

Marker plates on which authentic prostaglandins can be located by spraying subsequently with 10% phosphomolybdic acid are run simultaneously with preparative plates. Zones from the latter corresponding to the Rf value of the appropriate prostaglandin are then scraped off and the prostaglandin eluted from the silica gel with methanol. Alternatively, the marker prostaglandins may be run at the sides and the unknown at the middle of the same plate. After development the central portion of the plate is separated with a glass

Table 2. *Thin-layer chromatographic systems for the separation of prostaglandins*

Compound investigated	Developing solvent system composition (v/v)		Adsorbent	Detection method	Reference
PGE$_1$, PGF$_{1\alpha}$, PGF$_{1\beta}$ free acids	Chloroform Methanol Water	60 10 0.5	Kieselgel G in 0.2 N oxalic acid, plates pre-washed with develop-ing solvent	Acid ceric sulfate, heat and visualize with UV light	1
PGE$_1$, PGE$_2$, PGE$_3$, PGF$_{1\alpha}$, PGF$_{1\beta}$, PGF$_{2\alpha}$, PGF$_{2\beta}$, PGF$_{3\alpha}$, PGF$_{3\beta}$ free acids	Benzene Dioxane Acetic acid	20 20 1	Silica Gel G	10% Phosphomolybdic acid in ethanol, heat; sulphuric acid spray, heat	2, 3, 4, 5
	Ethyl acetate Acetic acid Methanol 2,2,4-Tri-methyl pentane Water (upper phase)	100 30 35 10 100	Silver nitrate (3—20%) impregnated Silica Gel G	10% Phosphomolybdic acid in ethanol, heat	
PGE$_1$, PGE$_2$, PGE$_3$ free acids	Ethyl acetate Acetic acid 2,2,4-Tri-methyl pentane Water (upper phase)	110 20 30 100	Silver nitrate (3—20%) impregnated Silica Gel G	2,4-Dinitrophenyl hydrazine; quantita-tion by alkaline iso-merization of TLC eluates	6, 7

Compound	Solvent system		Adsorbent	Detection	Ref.
Rat urinary metabolite of PGF$_{1\alpha}$ (2,3-dinorprostaglandin F$_{1\alpha}$) 19-Hydroxy-PGA$_1$, 19-Hydroxy-PGA$_2$	Ethyl acetate Acetic acid Methanol 2,2,4-Tri-methyl pentane Water (upper phase)	110 10 15 10 100	Silver nitrate impregnated Silica Gel G	10% Phosphomolybdic acid, heat	8
Guinea pig lung metabolites of PGE$_2$ (11α,15-dihydroxy-9-oxoprost-5-enoic and 11α-hydroxy-9,15-dioxoprost-5-enoic acids), PGF$_{2\alpha}$	Ethyl acetate Acetic acid 2,2,4-Tri-methyl pentane Water (upper phase)	100 10 30 100	Silica Gel G—silver nitrate (25 : 1)		9, 5
	Benzene Dioxane Acetic acid	65 15 2	Silica Gel G		
	Ethyl acetate Acetic acid 2,2,4-Tri-methyl pentane Water (upper phase)	110 20 30 100	Silica Gel G—silver nitrate (30 : 1)		

1 EGLINTON et al. (1963)
2 GRÉEN and SAMUELSSON (1964)
3 VAN DORP et al. (1964)
4 SAMUELSSON (1964 a)
5 ANGGÅRD and SAMUELSSON (1965 a)

6 BYGDEMAN and SAMUELSSON (1964)
7 BYGDEMAN and SAMUELSSON (1966)
8 GRANSTRÖM et al. (1965)
9 ÄNGGÅRD et al. (1965)
10 ÄNGGÅRD and SAMUELSSON (1965 b)

11 HAMBERG and SAMUELSSON (1966)
12 SCHNEIDER et al. (1966)
13 HORTON and THOMPSON (1964)

Table 2 (continued)

Compound investigated	Developing solvent system composition (v/v)		Adsorbent	Detection method	Reference
PGE_3 free acid	Ethyl acetate	110	Silica Gel G—silver nitrate (30 : 1)	Concentrated sulphuric acid or phosphomolyb-dic acid (10%) in ethanol, heat	10
	Acetic acid	15			
	2,2,4-Tri-methyl pentane	30			
	Water	100			
	(upper phase)				
PGB_1 and PGB_2 free acids	Ethyl acetate	90	Silica Gel G with and without silver nitrate		11
	Acetic acid	20			
	2,2,4-Tri-methyl pentane	50			
	Water	100			
	(upper phase)				
C^{14}-labelled PGE_1 and PGA_1 free acids	Methanol	5	Silica Gel	X-ray film	12
	Acetic acid	5			
	Chloroform	90			
PGE_1 free acid	Diethyl ether	50	Silica Gel G	Fluorescein, bromine, visualized under UV light	13
	Heavy petroleum	40			
	Acetic acid	5			
	Methanol	5			
PGE_1, $PGF_{1\alpha}$, $PGF_{1\beta}$ methyl esters	Chloroform	60	Kieselgel G	Iodine spray	1
	Methanol	10			
	Water	0.5			

Compound	Solvent system	Adsorbent	Detection	Ref.
PGE$_1$, PGE$_2$, PGE$_3$, PGF$_{1\alpha}$, PGF$_{1\beta}$, PGF$_{2\alpha}$, PGF$_{2\beta}$, PGF$_{3\alpha}$, PGF$_{3\beta}$, PGB$_1$, PGB$_2$, PGB$_3$ methyl esters	Benzene 5 Dioxan 4	Silica Gel G	10% Phosphomolybdic acid in ethanol, heat; sulphuric acid spray, heat	2, 5
	Ethyl acetate 8 Methanol 2 Water 5 (upper phase)	Silver nitrate (3—20%) impregnated Silica Gel G	10% Phosphomolybdic acid in ethanol, heat	
	Ethyl acetate 16 Methanol 2.5 Water 10 (upper phase)	Silver nitrate (3—20%) impregnated Silica Gel G	10% Phosphomolybdic acid in ethanol, heat	
PGE$_1$, PGF$_{1\alpha}$, PGF$_{1\beta}$ methyl ether methyl esters	Hexane 1 Diethyl ether 1 Chloroform	Kieselgel G Kieselgel G	Iodine spray Iodine spray	1
PGA$_1$ and PGA$_2$ acetylated methyl esters	Ethyl acetate 50 2,2,4-Tri-methyl pentane 100 Water 100 (upper phase)	Silica Gel G with and without silver nitrate		11
2,4-Dinitrophenyl hydrazones of reductive ozonolysis products of PGE$_3$ guinea pig lung metabolites	Diisopropyl ether 40 n-pentane 40	Silica Gel G		10

Table 2 (continued)

Compound investigated	Developing solvent system composition (v/v)		Adsorbent	Detection method	Reference
Reduced methyl ester of PGE$_3$ guinea pig lung metabolites (11α,15-dihydroxy-9-oxo-prosta-5,17-dienoic and 11α-hydroxy-9,15-dioxoprosta-5,17-dienoic acids)	Ethyl acetate Methanol Water (upper phase)	100 5 100	Silica Gel G—silver nitrate (30 : 1)	Concentrated sulphuric acid or phosphomolyb-dic acid (10%) in ethanol spray, heat	
19-Hydroxy-PGB$_1$, 19-Hydroxy-PGA$_2$, 19-Hydroxy-PGB$_2$	Ethyl acetate Acetic acid 2,2,4-Tri-methyl pentane Water (upper phase)	110 20 30 100	Silica Gel G		11

cutter and the two side portions sprayed — the appropriate zones on the central portion to be eluted are thus easily located.

It is preferable, however, to use radio-active tracers. A small quantity of labelled marker is added to the unknown — the zone(s) containing the radioactivity is then scraped off and eluted.

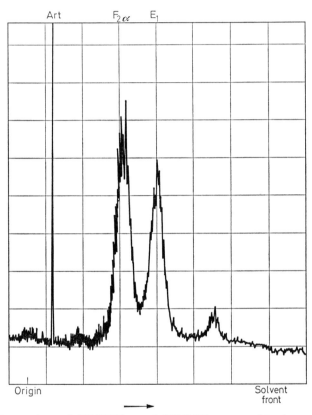

Fig. 7. Separation of ^{3}H-PGF$_{2\alpha}$ and ^{3}H-PGE$_1$ on a thin layer plate of silica gel G. Ordinate: cps; Abscissa: length of the plate

Thin layer chromatography using the MI system of GRÉEN and SAMUELSSON (1964) is a particularly convenient method for checking that methylation of a sample has gone to completion, since the free acid remains at the origin. The solvent system described by HORTON and THOMPSON (1964) is useful for separating prostaglandins from less polar lipids.

Many useful variations in technique have been described. These include the addition of $AgNO_3$ to the silica gel to effect the separation of prostaglandins which differ only in their degree of unsaturation, the use of repetitive development of the plate to improve separation and the preparative use of thick plates. Technical details may be found in the various papers cited, particularly those of the Swedish workers.

g) Gas-liquid Chromatography

Much of the early work on the gas chromatographic behaviour of the prostaglandins has been summarised in the excellent review by RAMWELL et al. (1968). It can be concluded that the best results are obtained if the hydroxyl and ketonic substituents of the prostaglandins are protected by appropriate derivative formation, in addition to esterification.

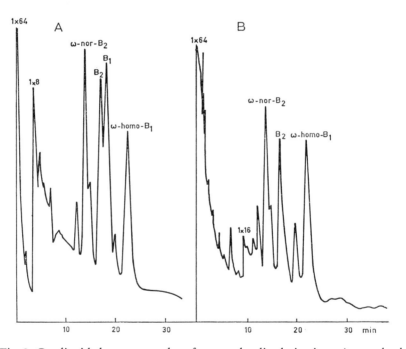

Fig. 8. Gas-liquid chromatography of prostaglandin dericatives. A, standard mixture, 10 ng each; B, analysis of rat epidermis; the ω-nor and ω-homo compounds were added as internal standards. (JOUVENAZ et al., 1970)

BYGDEMAN and SAMUELSSON (1966) used the trimethyl-silyl ether derivatives for separation and estimation of PGF compounds in human semen. Using a flame ionisation detector, amounts down to 50—100 ng could be estimated.

Greatly improved sensitivity has been achieved by JOUVENAZ, NUGTEREN, BEERTHIUS and VAN DORP (1970) using an electron capture detector. Although detection is restricted to PGB compounds which are esterified (with diazomethane) and silylated (with bis-trimethyl-silylacetamide), amounts down to 1 ng can be estimated (Fig. 8). Internal standards of ω-homo-E_1 and ω-nor-E_2 are added at the beginning of the extraction procedure. PGE compounds in the mixture are converted to the corresponding PGB by treatment with methanolic KOH. This method has the advantage of great sensitivity but is restricted to PGB compounds or to prostaglandins which can be converted to PGB (PGE and PGA). Identification by this method alone is inconclusive, since the occurrence of a peak corresponding to the retention volume or carbon value of a particular prostaglandin derivative is insufficient evidence. A similar criticism applies to the method of ALBRO and FISHBEIN (1969), although their method can be used for several different prostaglandins.

h) Radio Gas Chromatography

This is an indispensable tool for the study of prostaglandin metabolism. A typical experiment might be made as follows: A mixture containing radio-active (^3H or ^{14}C) prostaglandin and a larger amount of cold (unlabelled) prostaglandin would be incubated with an enzyme system or injected into an animal. Purification of the resulting metabolites by solvent extraction and partition chromatography would be monitored by liquid scintillation counting. Before identification of these unknown compounds could be attempted by combined gas chromatography-mass spectrometry, their gas chromatographic behaviour must be determined. This can be achieved by preparing appropriate derivatives and measuring their carbon values by a radio gas chromatograph, which records the amount of tritium of carbon-14 in the effluent from the column. A simultaneous record of total mass (cold and labelled) in the sample is obtained by use of a conventional detector (e. g. flame ionisation) (Fig. 9).

By this means it can be established whether a particular derivative of the metabolite(s) will chromatograph and if so its carbon value can be estimated as follows: mixtures of methyl esters of straight chain saturated fatty acids are used as standards and graphs are constructed by plotting the logarithm of the observed retention time as a function of the number of carbon atoms of the esters on a linear

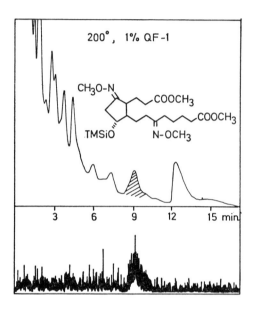

Fig. 9. Radio gas chromatography (lower trace) and gas chromatography of a mixture containing the main human urinary metabolite of ³H-PGE₂ (and PGE₂) run as the methoxime-methyl ester-trimethylsilylether (SAMUELSSON et al., 1971)

scale. Retention times of the prostaglandin derivatives and of un-known metabolites are converted to carbon values by reference to the diagram (HAMBERG, 1968).

This technique applied to different derivatives of a radio-active metabolite of unknown structure can give useful preliminary evidence of identification. Conclusive evidence can usually be obtained by analysing the same derivative by combined gas chromatography mass spectrometry. Mass spectra are taken at times corresponding to the carbon values found by radio gas chromatography.

3. Identification and Estimation of Prostaglandins

a) Simple Tests of Identification

If an unknown smooth muscle stimulating or depressor substrance is a prostaglandin, it will be extractable into ethyl acetate from an acidified aqueous medium. Further partition will show that it behaves like a polar acidic lipid.

If the unknown is a prostaglandin, its biological activity will be greatly reduced by incubation with a preparation of the enzyme, 15-hydroxy-prostaglandin dehydrogenase (ÄNGGÅRD, 1968). This enzyme is specific for prostaglandins and should be useful for distinguishing between prostaglandins and other naturally occurring substances which affect smooth muscle.

Girard's reagent T (Trimethylammonium-acetohydrazine chloride) inactivates PGE but not PGF compounds (AMBACHE and BRUMMER, 1968). Similarly PGE compounds are biologically inactivated by treatment with alcoholic potassium hydroxide by conversion to the corresponding PGB. The formation of PGB may be detected by ultraviolet absorption at 278 nm.

b) Parallel Biological Assay

Biological preparations have been used extensively for the detection and estimation of prostaglandins because of their great sensitivity (Table 3). Prostaglandins of the A, E and F series can be distinguished readily by parallel assay on appropriate tissues. Quantitative parallel assay is a powerful tool; its value should not be underestimated particularly if results can be obtained on four or more tissues. Thus

Table 3. *Amounts (ng) of prostaglandins which can be estimated by biological assay*

Preparation	E_1	E_2	$F_2\alpha$	A_2
Rat fundus	5	2	5	100
Cat B. P.	100	50	5000	5
Guinea-pig ileum	5	5	20	
Rabbit jejunum	50	50	2	
Cat trachea	1	1	50	

a PGE can be distinguished from a PGF by estimating the unknown in terms of a compound of both series on rabbit jejunum and guinea-pig ileum (or cat trachea) (HORTON and MAIN, 1963, 1965). The cat blood pressure and rat fundus combination would easily distinguish a PGA compound from either a PGE or PGF (HORTON and JONES, 1969). It has not proved possible to distinguish between prostaglandins of the same series for example PGE_1 and PGE_2, chromatographic or other evidence is needed.

An extension of parallel biological assay which may obviate the need for preliminary extraction is the cascade of superfused tissues developed by VANE (1969). Blood or perfusion fluid from an animal or an organ is allowed to drip over the surface of smooth muscle preparations selected for their sensitivity to various endogenous substances. By recording the contractions or relaxations of six or more tissues arranged in series or in parallel it is possible to measure changes in levels of circulating hormones or the output of hormones from tissues. Greater specificity is obtained by the use of specific blocking drugs. Concentrations are estimated by comparing responses with those produced by authentic prostaglandins added to the bathing fluid.

By this method it is possible to obtain information very rapidly about the release and disappearance of pharmacologically-active compounds. On the other hand, this method does not provide conclusive evidence of identification and if more than one prostaglandin is present accurate estimation is very difficult. It would, for example, be impossible to measure the amounts of $PGF_{1\alpha}$ and $PGF_{2\alpha}$ if they were present in a mixture.

All biological assay methods suffer from the great disadvantage that they cannot be used to measure inactive metabolites. The level of metabolites in the circulation may of course be as important as that of the parent compound. For such work, other methods are needed.

c) Enzymatic Determination

Using a 15-hydroxy-prostaglandin dehydrogenase preparation from swine lung, a very sensitive method which is specific for prostaglandins hydroxylated in the 15 position, has been evolved (ÄNG-GÅRD, 1968). In spite of its specificity for prostaglandins and of its very high sensitivity, this method has not so far been widely used.

This is partly because the enzyme is difficult to prepare but also because the different prostaglandins must first be separated before estimations can be made. The method provides no evidence of identification other than that the compound is a 15-hydroxy-prostaglandin. In this respect it may be considered less specific than parallel biological assay.

d) Radio-immunoassay and Competitive Protein Binding

These techniques are in the early stages of development with respect to prostaglandins, but it can be anticipated by analogy with the steroid field that such methods may in the future form the basis for routine estimations of prostaglandin levels in body fluids. They have the advantage of simplicity and high sensitivity. A note of caution however must be sounded about their specificity. It would not be justifiable to claim that a method for the estimation of a particular prostaglandin (say PGE_2) is specific on the grounds that the antibody (or other protein) does not bind $PGF_{2\alpha}$ and PGA_2. The numerous metabolites and other natural analogues of PGE_2 (see Chapter 5) must also be tested. Furthermore, the presence of as yet unidentified prostaglandins in the sample must always remain a possibility. It would be advisable in any series of estimations by radio-immunoassay that the identity of the prostaglandin being estimated should be confirmed by combined gas chromatography and mass spectrometry (see below).

e) Ultraviolet, Infra-red and Nuclear Magnetic Resonance Spectroscopy

These methods were used extensively in elucidating the structure of the prostaglandins. I.R. and N.M.R. give valuable information provided that milligram quantities are available for analysis, but they have little place in the identification of prostaglandins in the microgram or nanogram range, the amounts likely to be found in small quantities of tissues or tissue fluids. Absorption of PGA and PGB compounds at 217 and 278 nm respectively is useful, the latter has been applied to the assay of the enzyme, prostaglandin isomerase (see Chapter V) and the estimation of PGE, PGA and PGB levels (in terms of PGB) in human seminal plasma (see Chapter XV).

f) Mass Spectrometry

BERGSTRÖM, SAMUELSSON their co-workers have made extensive use of mass spectrometry for the identification of prostaglandins and their metabolites in animal tissues. By linking a mass spectrometer to a gas chromatograph (as in the LKB 9000) they have obtained conclusive evidence of identification of substances eluted from the column. This powerful combination is indispensable to the future development of the prostaglandin field.

Perhaps the most important application of combined gas chromatography-mass spectrometry is in the identification and estimation of prostaglandins and their metabolites in body fluids, where the amounts available for analysis may be considerably less than one microgram. We have recently developed a micro-method for the identification and estimation of prostaglandins in the nanogram range (THOMPSON, LOS and HORTON, 1970).

All samples are methylated by reaction for 20 min with a freshly prepared solution of diazomethane in diethyl ether-methanol (9 : 1). Trimethylsilyl ethers are formed by the addition of 25 μl bistri-methylsilyl trifluoroacetamide (BSTFA) to the ester, 10 μl samples are injected on to the column after 3 hours without removal of the BSTFA. Alternatively trifluoroacetates are prepared by reacting the methyl esters for 2 hours with 200 μl trifluoroacetic anhydride which is then removed in a vacuum desiccator. The residue is dissolved in 25 μl hexane, 10 μl being injected. Analyses are performed on an LKB 9000 gas chromatograph-mass spectrometer. The column (1.5 m $\times$ 1.5 mm i. d.) is packed with 3% OV1 on Supasorb AW, 100—200 mesh pre-treated with dimethylchlorosilane in carbon tetrachloride. The carrier gas, helium, flows at 20 ml/min. Mass spectra are recorded at an electron voltage of 27.5.

PGA_1, PGA_2, PGB_1 and PGB_2 which co-chromatograph on silicic acid and Sephadex LH20 can be separated from each other as their methyl ester-trimethyl silyl ethers by gas chromatography and their identity established by a mass spectrum of the column effluent taken at the peak of the gas chromatographic trace. The trifluoro-acetates of PGA_1 and PGA_2 chromatograph on our system but the corresponding derivatives of PGB_1 and PGB_2 do not.

PGE_1 and PGE_2 can be chromatographed on the nanogram scale as their methoxime-trimethylsilyl ether-methyl ester derivatives (each compound showing two peaks corresponding to the two epimers formed). The trifluoroacetates of PGE_1 and PGE_2 (which are dehydrated to the corresponding PGA compound) can also be separated.

$PGF_{1\alpha}$ and $PGF_{2\alpha}$ together with the 19-hydroxy derivatives of PGA and PGB can be separated by this system either as the trimethylsilyl ether-methyl esters or the trifluoroacetates.

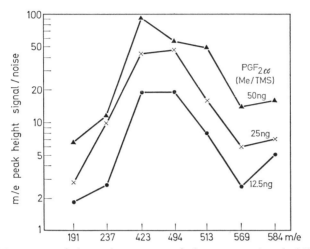

Fig. 10. Mass spectral data of $PGF_{2\alpha}$ (methyl ester — trimethylsilyl ether) 50, 25 and 12.5 ng injected on the gas chromatographic column, mass spectra taken at the retention time corresponding to the peak maximum on the GC trace. Abscissa: m/e value, Ordinate: signal to noise ratio. (HORTON et al., 1971)

Mass spectra of all these prostaglandins can be obtained with amounts (injected) down to about 10 ng (Fig. 10). Identification of an unknown is based on a mass spectral-gas chromatographic comparison with an authentic sample of the particular prostaglandin or metabolite.

The lower limit of sensitivity is partly determined by the amount of data considered essential for conclusive identification. If it is desired to see all the characteristic peaks of a particular prostaglandin derivative, then the lower limit is unlikely to be less than 10 ng and with partially purified extracts of biological origin the lower limit may be as high as 50 ng. On the other hand if the occurrence of one or two of the most abundant m/e peaks in the spectrum at the retention time corresponding to the carbon value of the derivative being studied is regarded as sufficient evidence, then far greater sensitivity can be achieved.

By mass fragmentography in which a particularly abundant peak is selected (for example m/e 314 for the TFA of MePGF$_{2\alpha}$) it is possible to detect far less than 10 ng. Similar results can be obtained using

Table 4. *Release of prostaglandins from tissues*

Site	Stimulus	Reference
Cat superfused somatosensory cortex	Spontaneous Sensory nerve stimulation Contralateral cortical stimulation	1, 2, 3, 21, 4, 26
Cat superfused cerebellar cortex	Spontaneous	6, 7
Cat perfused cerebral ventricles	Spontaneous	8
Dog perfused cerebral ventricles	Spontaneous 5-Hydroxytryptamine	9
Frog superfused spinal cord	Spontaneous Sensory nerve stimulation	10, 3
Rat phrenic nerve-diaphragm *in vitro*	Electrical stimulation Noradrenaline	11, 28
Rat epididymal fat pad *in vitro*	Nerve stimulation Noradrenaline	3
Rat gastro-intestinal tract	Pentagastrin Histamine Vagal stimulation Transmural stimulation Carbachol 5-Hydroxytryptamine	12, 13, 14 15, 32
Dog blood-perfused spleen	Nerve stimulation Adrenaline Colloidal particles	16, 17, 18 30
Cat Ringer-perfused adrenals	Acetylcholine	5, 11
Guinea-pig Ringer-perfused lungs	Phospholipase A	19
Dog Ringer-perfused lungs	Stretch	29
Rabbit eye	Mechanical stimulation	20
Frog intestine *in vitro*	Spontaneous	22, 23, 24
Human medullary carcinoma of the thyroid *in vivo*	Spontaneous	25
Rat carrageenin pouch	Inflammatory response	27
Dog kidney	Ischaemia	31
Guinea-pig uterus	Distension, oestrogen	33, 34

multiple ion detection, in which three peaks (fairly closely spaced in the spectrum) are selected. The evidence of identification using these methods depends upon finding one, two or three peaks at the retention time corresponding to that of an authentic sample of the prostaglandin derivative. The chances of finding another compound which combines three m/e peaks at the same ratio with identical gas chromatographic characteristics must be exceedingly small. Nevertheless, the evidence is less conclusive than if a full spectrum of the unknown is obtained. The choice between mass fragmentography (or multiple ion detection) and mass spectrometry can only be made in the light of the problem being investigated.

There are numerous substances which may interfere with good gas chromatography-mass spectrometry. Compounds used in cleaning glassware, traces of silicone grease, impurities from the reagents used in the extraction procedure, blood pigments and other polar lipids must all be kept to a minimum (preferably excluded) when working in the low nanogram range.

Another problem is that of quantitation at these low levels. Losses on the gas chromatograph column and in the separator are variable but may be large. A solution to this may have been achieved by SAMUELSSON, HAMBERG and SWEELEY (1970) using a reversed isotope dilution technique. To nanogram quantities of the extracted PGE_1

---

1 RAMWELL and SHAW (1963 a)
2 RAMWELL and SHAW (1963 b)
3 RAMWELL and SHAW (1967)
4 RAMWELL et al. (1966 a)
5 SHAW and RAMWELL (1967)
6 COCEANI and WOLFE (1965)
7 WOLFE et al. (1967)
8 FELDBERG and MYERS (1966)
9 HOLMES (1970)
10 RAMWELL et al. (1966 b)
11 RAMWELL et al. (1965)
12 SHAW and RAMWELL (1968)
13 BENNETT et al. (1967)
14 COCEANI et al. (1967)
15 COCEANI et al. (1968)
16 DAVIES et al. (1968)
17 FERREIRA and VANE (1967)
18 GILMORE, VANE and WYLLIE (1969)
19 BABILLI and VOGT (1965)
20 AMBACHE et al. (1965)
21 RAMWELL and SHAW (1966)
22 BARTELS et al. (1968)
23 SUZUKI and VOGT (1965)
24 VOGT and DISTELKÖTTER (1967)
25 WILLIAMS et al. (1968)
26 BRADLEY et al. (1969)
27 WILLIS (1969)
28 LAITY (1969)
29 EDMONDS et al. (1969)
30 PIPER and VANE (1969)
31 CROWSHAW et al. (1969)
32 RADMANOVIĆ (1968)
33 POYSER et al. (1970, 1971)
34 BLATCHLEY et al. (1971)

(as the methoxime-methyl ester) to be estimated, is added a microgram of the deuterated methoxime of methyl PGE_1. These two derivatives co-chromatograph and so the excess of the deuterated derivative acts as a carrier for the natural prostaglandin both through the column and the separator. Furthermore, by scanning for abundant m/e values and the corresponding m/e + 3 in the case of the deuterated material, the ratio of unknown to added PGE_1 can be determined. Since the amount of added PGE_1 is known, the amount of unknown can be estimated. So far this method has been applied only to PGE compounds. The deuterated derivative must obviously contain a very high percentage of deuterium since any non-deuterated material added will be estimated as if it were the unknown.

4. Occurrence and Release of Prostaglandins

In the light of the methods described above and taking into account the large number of closely related prostaglandins and metabolites, the evidence for the occurrence of prostaglandins in tissues and their release on stimulation may be evaluated critically.

Some of the reports of identification although not based upon conclusive evidence contain such detailed evidence of various kinds that the identification of the particular prostaglandin is hardly in doubt. In other instances the evidence could really only justify the term "prostaglandin-like" or possibly for example, "PGE-like". In spite of the relative paucity of conclusive evidence, it is apparent that substances of the prostaglandin type occur in almost every mammalian tissue (KARIM, HILLIER and DEVLIN, 1968; KARIM, SANDLER and WILLIAMS, 1967). They are present in all regions of the brain (HORTON and MAIN, 1967; HOLMES and HORTON, 1968). They have been detected in several subcellular fractions; for example, in whole rabbit brain homogenates $PGF_{2\alpha}$ is predominantly in the cytoplasmic fraction (HOPKIN, HORTON and WHITTAKER, 1968) but in rat cortical tissue, a high proportion of the prostaglandins are associated with the mitochondrial and synaptosomal fractions (KATAOKA, RAMWELL and JESSUP, 1967).

Release of prostaglandins or prostaglandin-like substances from numerous tissues and organs has been reported (Table 4). In a few instances gas chromatographic-mass spectrometric identification of the

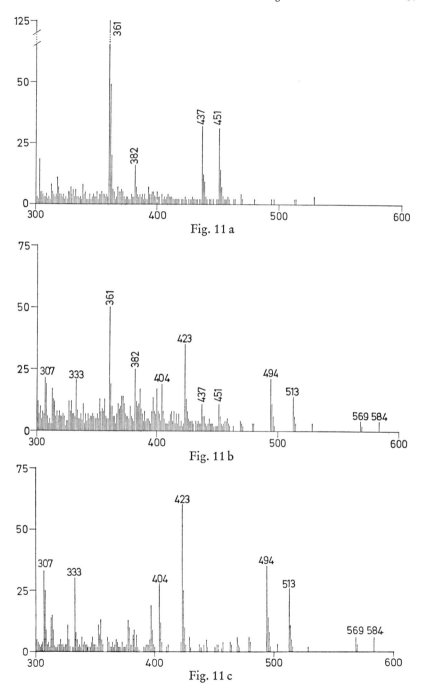

Fig. 11 a

Fig. 11 b

Fig. 11 c

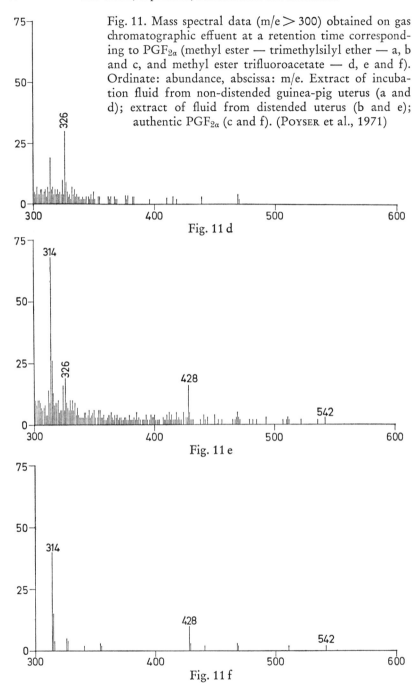

Fig. 11. Mass spectral data (m/e > 300) obtained on gas chromatographic effluent at a retention time corresponding to PGF$_{2\alpha}$ (methyl ester — trimethylsilyl ether — a, b and c, and methyl ester trifluoroacetate — d, e and f). Ordinate: abundance, abscissa: m/e. Extract of incubation fluid from non-distended guinea-pig uterus (a and d); extract of fluid from distended uterus (b and e); authentic PGF$_{2\alpha}$ (c and f). (POYSER et al., 1971)

Fig. 11 d

Fig. 11 e

Fig. 11 f

prostaglandin has been achieved, for example PGE$_2$ and PGF $_{2\alpha}$ from the dog spleen (BEDWANI, HORTON, MILLAR and THOMPSON, unpublished), PGF$_{2\alpha}$ from the cat superior cervical ganglion (DAVIS, HORTON, JONES and QUILLIAM, 1971), and PGF$_{2\alpha}$ from the guinea-pig and sheep uterus (POYSER, HORTON, THOMPSON and LOS, 1970, 1971; BLATCHLEY, DONOVAN, POYSER, HORTON, THOMPSON and LOS, 1971; BLAND, HORTON and POYSER, 1971; HORTON, JONES, POYSER and THOMPSON, 1971) (Fig. 11). Similar confirmation is needed with respect to prostaglandin output from other sources since adequate identification is a prerequisite for proper quantitation.

References

ALBRO, P. W., FISHBEIN, L.: Determination of prostaglandins by gas-liquid chromatography. J. Chromatog. **44**, 443—451 (1969).

AMBACHE, N., BRUMMER, H. C.: A simple chemical procedure for distinguishing E from F prostaglandins, with application to tissue extracts. Br. J. Pharmac. **33**, 162—170 (1968).

— KAVANAGH, L., WHITING, J.: Effect of mechanical stimulation on rabbits' eyes: release of active substance in anterior chamber perfusates. J. Physiol. (Lond.) **176**, 378—408 (1965).

ÄNGGÅRD, E.: The isolation and determination of prostaglandins in lungs of sheep, guinea-pig, monkey and man. Biochem. Pharmac. **14**, 1507—1516 (1965).

— BERGKVIST, H.: Group separation of prostaglandins on Sephadex LH-20. J. Chromatog. **48**, 542—544 (1970).

— GRÉEN, K., SAMUELSSON, B.: Synthesis of tritium labeled prostaglandin E$_2$ and studies on its metabolism in guinea-pig lung. J. biol. Chem. **240**, 1932—1940 (1965).

— SAMUELSSON, B.: Smooth muscle stimulating lipids in sheep iris. The identification of prostaglandin F$_{2\alpha}$. Biochem. Pharmac. **13**, 281—283 (1964).

— — Biosynthesis of prostaglandins from arachidonic acid in guinea pig lung. J. biol. Chem. **240**, 3518—3521 (1965 a).

— — The metabolism of prostaglandin E$_3$ in guinea pig lung. Biochemistry, N. Y. **4**, 1864—1871 (1965 b).

— — Purification and properties of a 15-hydroxy prostaglandin dehydrogenase from swing lung. Ark. Kemi, **25**, 293—300 (1966).

BABILLI, S., VOGT, W.: Nature of the fatty acids acting as 'slow reacting substance' (SRS-C). J. Physiol., Lond. **177**, 31 P—32 P (1965).

BARTELS, J., VOGT, W., WILLE, G.: Prostaglandin release from and formation in perfused frog intestine. Arch. Pharmak. exp. Path. **259**, 153 to 154 (1968) (see erratum **259**, 459 (1968)).

BENNETT, A., FRIEDMANN, C. A., VANE, J. R.: Release of prostaglandin E_1 from the rat stomach. Nature **216**, 873—876 (1967).

BERGSTRÖM, S., DRESSLER, F., KRABISCH, L., RYHAGE, R., SJÖVALL, J.: The isolation and structure of a smooth muscle stimulating factor in normal sheep and pig lungs. Ark. Kemi **20**, 63—66 (1962).

— — RYHAGE, R., SAMUELSSON, B., SJÖVALL, J.: The isolation of two further prostaglandins from sheep prostate glands. Ark. Kemi **19**, 563—567 (1962).

— KRABISCH, L., SJÖVALL, J.: Smooth muscle stimulating factors in ram semen. Acta chem. scand. **14**, 1706—1710 (1960).

— SAMUELSSON, B.: Isolation of prostaglandin E_1 from human seminal plasma. J. biol. Chem. **237**, PC 3005—3006 (1962).

— — Isolation of prostaglandin E_1 from calf thymus. Acta chem. scand. **17**, 282—287 (1963).

— SJÖVALL, J.: The isolation of prostaglandin F from sheep prostate glands. Acta chem. scand. **14**, 1693—1700 (1960 a).

— — The isolation of prostaglandin E from sheep prostate glands. Acta chem. scand. **14**, 1701—1705 (1960 b).

BLAND, K. P., HORTON, E. W., POYSER, N. L.: Levels of prostaglandin F_{2a} in uterine venous blood of sheep during the oestrous cycle. Life Sci. **10**, 509—517 (1971).

BLATCHLEY, F. R., DONOVAN, B. T., POYSER, N. L., HORTON, E. W., THOMPSON, C. J., LOS, M.: Identification of prostaglandin F_{2a} in the utero-ovarian blood of guinea-pig after treatment with oestrogen. Nature, **230**, 243—244 (1971).

BRADLEY, P. B., SAMUELS, G. M. R., SHAW, J. E.: Correlation of prostaglandin release from the cerebral cortex of cats with the electrocorticogram, following stimulation of the reticular formation. Br. J. Pharmac. **37**, 151—157 (1969).

BYGDEMAN, M., HOLMBERG, O.: Isolation and identification of prostaglandins from ram seminal plasma. Acta chem. scand. **20**, 2308—2310 (1966).

— SAMUELSSON, B.: Quantitative determination of prostaglandins in human semen. Clinica chim. Acta **10**, 566—568 (1964).

— — Analyses of prostaglandins in human semen. Clinica chim. Acta **13**, 465—474 (1966).

COCEANI, F., PACE-ASCIAK, C., VOLTA, F., WOLFE, L. S.: Effect of nerve stimulation on prostaglandin formation and release from the rat stomach. Am. J. Physiol. **213**, 1056—1064 (1967).

— — WOLFE, L. S.: Studies on the effect of nerve stimulation on prostaglandin formation and release in the rat stomach. Prostaglandin Symposium of the Worcester Foundation for Exp. Biol., Ed. P. W. RAMWELL and J. E. SHAW, pp. 39—45, Interscience, New York 1968.

— WOLFE, L. S.: Prostaglandins in brain and the release of prostaglandin-like compounds from the cat cerebellar cortex. Can. J. Physiol. Pharmac. **43**, 445—450 (1965).

CROWSHAW, K., McGIFF, J. C., TERRAGNO, N. A., LONIGRO, A. J., WILLIAMSON, M. A., STRAND, J. C., LEE, J. B., NG, K. K. F.: Prostaglandin-like substances present in blood during renal ischemia: Patterns of release and their partial characterization. J. Lab. clin. Med. 74, 866 (1969).

DANIELS, E. G., HINMAN, J. W., LEACH, B. E., MUIRHEAD, E. E.: Identification of prostaglandin E_2 as the principal vaso-depressor lipid of rabbit renal medulla. Nature 215, 1298—1299 (1967).

— PIKE, J. E.: Isolation of prostaglandins. Prostaglandin Symposium of the Worcester Foundation for Exp. Biol. Ed.: P. W. RAMWELL and J. E. SHAW, pp. 379—387, Interscience, New York 1968.

DAVIES, B. N., HORTON, E. W., WITHRINGTON, P. G.: The occurrence of prostaglandin E_2 in splenic venous blood of the dog following splenic nerve stimulation. Br. J. Pharmac. 32, 127—135 (1968).

DAVIS, H. A., HORTON, E. W., JONES, K. B., QUILLIAM, J. P.: The identification of prostaglandins in pre-vertebral venous blood after pre-ganglionic stimulation of the cat superior cervical ganglion. Br. J. Pharmac. 42, 569—583 (1971).

EDMONDS, J. F., BERRY, E., WYLLIE, J. H.: Release of prostaglandins caused by distension of the lungs. Br. J. Surg. 56, 622—623 (1969).

EGLINTON, G., RAPHAEL, R. A., SMITH, G. N., HALL, W. J., PICKLES, V. R.: The isolation and identification of two smooth muscle stimulants from menstrual fluid. Nature 200, 960, 993—995 (1963).

FELDBERG, W., MYERS, R. D.: Appearance of 5-hydroxytryptamine and an unidentified pharmacologically active lipid acid in effluent from perfused cerebral ventricles. J. Physiol., Lond. 184, 837—855 (1966).

FERREIRA, S. H., VANE, J. R.: Prostaglandins: their disappearance from and release into the circulation. Nature 216, 868—873 (1967).

GILMORE, N., VANE, J. R., WYLLIE, J. H.: Prostaglandin released by the spleen. Nature 218, 1135—1140 (1968).

— — — Prostaglandin released by the spleen in response to infusion of particles. In: Prostaglandins, Peptides and Amines. Ed.: P. MANTEGAZZA and E. W. HORTON. London: Academic Press 1969, pp. 21—29.

GRANSTRÖM, E., INGER, U., SAMUELSSON, B.: The structure of a urinary metabolite of prostaglandin $F_{1\alpha}$ in the rat. J. biol. Chem. 240, 457—461 (1965).

GRÉEN, K., SAMUELSSON, B.: Thin-layer chromatography of the prostaglandins. J. Lipid Res. 5, 117—120 (1964).

HAMBERG, M.: Metabolism of prostaglandins in rat liver mitochondria. Eur. J. Biochem. 6, 135—146 (1968).

— SAMUELSSON, B.: Prostaglandins in human seminal plasma. J. biol. Chem. 241, 257—263 (1966).

HICKLER, R. B.: The identification and measurement of prostaglandin in human plasma. Prostaglandin Symposium of the Worcester Foundation for Exp. Biol. Eed.: P. W. RAMWELL and J. E. SHAW. New York: Interscience 1968, pp. 279—293.

HOLMES, S. W.: The spontaneous release of prostaglandins into the cerebral ventricles of the dog and the effect of external factors on this release. Br. J. Pharmac. **38**, 653—658 (1970).
— HORTON, E. W.: The identification of four prostaglandins in dog brain and their regional distribution in the central nervous system. J. Physiol. Lond. **195**, 731—741 (1968).
— — STEWART, M. J.: Observations on the extraction of prostaglandins from blood. Life Sci. **7**, 349—354 (1968).
HOPKIN, J. M., HORTON, E. W., WHITTAKER, V. P.: Prostaglandin content of particulate and supernatant fractions of rabbit brain homogenates. Nature **217**, 71—72 ((1968).
HORTON, E. W., JONES, R. L.: Prostaglandins A_1, A_2 and 19-hydroxy-A_1; their actions on smooth muscle and their inactivation on passage through the pulmonary and hepatic portal vascular beds. Br. J. Pharmac. **37**, 705—722 (1969).
— — POYSER, N. L., THOMPSON, C. J.: The release of prostaglandins. Ann. N. Y. Acad. Sci. **180**, 351—361 (1971).
— MAIN, I. H. M.: A comparison of the biological activities of four pro-staglandins. Br. J. Pharmac. **21**, 182—189 (1963).
— — A comparison of the actions of prostaglandins $F_{2\alpha}$ and E_1 on smooth muscle. Br. J. Pharmac. **24**, 470—476 (1965).
— — Identification of prostaglandins in central nervous tissues of the cat and chicken. Br. J. Pharmac. **30**, 582—602 (1967).
— THOMPSON, C. J.: Thin-layer chromatography and bioassay of pro-staglandins in extracts of semen and tissues of the male reproductive tract. Br. J. Pharmac. **22**, 183—188 (1964).
JOUVENAZ, G. H., NUGTEREN, D. H., BEERTHUIS, R. K., VAN DORP, D. A.: A sensitive method for the determination of prostaglandins by gas chromatography with electron-capture detection. Biochim. biophys. Acta **202**, 231—234 (1970).
KARIM, S. M. M., HILLIER, K., DEVLIN, J.: Distribution of prostaglandins E_1, E_2, $F_{1\alpha}$ and $F_{2\alpha}$, in some animal tissues. J. Pharm. Pharmac **20**, 749—753 (1968).
— SANDLER, M., WILLIAMS, E. D.: Distribution of prostaglandins in human tissues. Br. J. Pharmac. **31**, 340—344 (1967).
KATAOKA, K., RAMWELL, P. W., JESSUP, S.: Prostaglandins: localization in subcellular particles of rat cerebral cortex. Science, N. Y. **157**, 1187—1189 (1967).
LAITY, J. L. H.: The release of prostaglandin E_1 from the rat phrenic nerve-diaphragm preparation. Br. J. Pharmac. **37**, 698—704 (1969).
LEE, J. B., CROWSHAW, K., TAKMAN, B. H., ATTREP, K. A., GOUGOUTAS, J. Z.: The identification of prostaglandins E_2, F_2 and A_2 from rabbit kidney medulla. Biochem. J. **105**, 1251—1260 (1967).
NORMAN, A., SJÖVALL, J.: On the transformation and enterohepatic circula-tion of cholic acid in the rat. J. biol. Chem. **233**, 872—885 (1958).

NUGTEREN, D. H., VONKEMAN, H., VAN DORP, D. A.: Non-enzymic conversion of *all cis* 8,11,14-eicosatrienoic acid into prostaglandin E_1. Recl. Trav. chim. Pays-Bas Belg. **86**, 1237—1245 (1967).

PIPER, P. J. and VANE, J. R.: The release of prostaglandins during anaphylaxis in guinea-pig isolated lungs. In: Prostaglandins, Peptides and Amines. Ed.: P. MANTEGAZZA and E. W. HORTON. London: Academic Press 1969, pp. 15—19.

POYSER, N. L., HORTON, E. W., THOMPSON, C. J., LOS, M.: Identification of prostaglandin $F_{2\alpha}$ released by distension of the guinea-pig uterus in vitro. J. Endocrin. **48**, xliii (1970).

— — — — Identification of prostaglandin $F_{2\alpha}$ released by distension of the guinea-pig uterus in vitro. Nature **230**, 526—528 (1971).

RADMANOVIĆ, B.: Prostaglandins in perfusate of the rat small intestine after vagal stimulation. Jugoslav. physiol. pharmac. Acta 4, 123—124 (1968).

RAMWELL, P. W., SHAW, J. E.: The nature of noncholinergic substances released from the cerebral cortex of cats on direct and indirect stimulation. J. Physiol. Lond. **169**, 51—52 P (1963 a).

— — The spontaneous and evoked release of noncholinergic substances from the cerebral cortex of cats. Life Sci. **2**, 419—426 (1963 b).

— — Spontaneous and evoked release of prostaglandins from the cerebral cortex of anesthetized cats. Am. J. Physiol. **211**, 125—134 (1966).

— — Prostaglandin release from tissue by drug, nerve and hormone stimulation. Nobel Symposium 2, Prostaglandins. Ed.: S. BERGSTRÖM and B. SAMUELSSON. Stockholm: Almqvist and Wiksell 1967, pp. 283—292.

— — CLARKE, G. B., GROSTIC, M. F., KAISER, D. G., PIKE, J. E.: Prostaglandins. In: Progress in the Chemistry of Fats and Other Lipids, Vol. 9. Ed.: R. T. HOLMAN. Oxford: Pergamon Press 1968, pp. 231—273.

— — DOUGLAS, W. W., POISNER, A. M.: Efflux of prostaglandin from adrenal glands stimulated with acetylcholine. Nature **210**, 273—274 (1966).

— — JESSUP, R.: Spontaneous and evoked release of prostaglandins from frog spinal cord. Am. J. Physiol. **211**, 998—1004 (1966).

— — KUCHARSKI, J.: Prostaglandin release from the rat phrenic nerve-diaphragm preparation. Science, N. Y. **149**, 1390—1391 (1965).

SAMUELSSON, B.: Isolation and identification of prostaglandins from human seminal plasma. J. biol. Chem. **238**, 3229—3234 (1963).

— Identification of a smooth muscle-stimulating factor in bovine brain. Biochim. biophys. Acta 84, 218—219 (1964 a).

— The identification of prostaglandin $F_{3\alpha}$ in bovine lung. Biochim. biophys. Acta 84, 707—713 (1964 b).

— HAMBERG, M., SWEELEY, C. C.: Quantitative gas chromatography of prostaglandin E_1 at the nanogram level. Anal. Biochem. **38**, 301—304 (1970).

SCHNEIDER, W. P., PIKE, J. E., KUPIECKI, F. P.: Determination of the origin of 9-keto-15-hydroxy-10,13-prostadienoic acid by a double-labeling technique. Biochim. biophys. Acta 125, 611—613 (1966).

SHAW, J. E., RAMWELL, P. W.: Prostaglandin release from the adrenal gland. Nobel Symposium 2, Prostaglandins. Ed.: S. BERGSTRÖM and B. SAMUELSSON. Stockholm: Almqvist and Wiksell 1967, pp. 293—299.

— — Release of prostaglandin from rat epididymal fat pad on nervous and hormonal stimulation. J. biol. Chem. 243, 1498—1503 (1968).

SUZUKI, T., VOGT, W.: Prostaglandine in einem Darmstoffpräparat aus Froschdarm. Arch. exp. Path. Pharmak. 252, 68—78 (1965).

THOMPSON, C. J., LOS, M., HORTON, E. W.: Separation, identification and estimation of prostaglandins in nanogram quantities by combined gas chromatography-mass spectrometry. Life Sci. 9, 983—988 (1970).

UNGER, W. G., STAMFORD, I. F., BENNETT, A., Extraction of prostaglandins from human blood. Nature 233, 336—337 (1971).

VAN DORP, D. A., BEERTHUIS, R. K., NUGTEREN, D. H., VONKEMAN, H.: Enzymatic conversion of all-cis-polyunsaturated fatty acids into prostaglandins. Nature 203, 839—841 (1964).

VANE, J. R.: The release and fate of vasoactive hormones in the circulation. Br. J. Pharmac. 35, 209—242 (1969).

VOGT, W., DISTELKÖTTER, B.: Release of prostaglandin from frog intestine. Nobel Symposium 2, Prostaglandins. Ed.: S. BERGSTRÖM and B. SAMUELSSON. Stockholm: Almqvist and Wiksell 1967, pp. 237—240.

WILLIAMS, E. D., KARIM, S. M. M., SANDLER, M.: Prostaglandin secretion by medullary carcinoma of the thyroid. Lancet 1, 22—23 (1968).

WILLIS, A. L.: Parallel assay of prostaglandin-like activity in rat inflammatory exudate by means of cascade superfusion. J. Pharm. Pharmac. 21, 126—128 (1969).

WOLFE, L. S., COCEANI, F., PACE-ASCIAK, C.: Brain prostaglandins and studies of the action of prostaglandins on the isolated rat stomach. Nobel Symposium 2, Prostaglandins. Ed.: S. BERGSTRÖM and B. SAMUELSSON. Stockholm: Almqvist and Wiksell 1967, pp. 265—275.

III. Synthesis

Total chemical synthesis of all the major natural prostaglandins is now possible as a result of the elegant methods devised by E. J. COREY. The synthesis illustrated in Fig. 1 gives high yields of the prostaglandins in their natural stereochemical configuration and may prove commercially feasible for large scale production.

From a common intermediary, 11,15-bis-tetrahydropyranyl ether of $PGF_{2\alpha}$, PGE_1, PGE_2, $PGF_{1\alpha}$ and $PGF_{2\alpha}$ can readily be obtained (COREY, NOYORI and SCHAAF, 1970). PGA_1, PGB_1, PGA_2 and PGB_2 can then be derived easily from PGE_1 and PGE_2. By slight modifications of the synthesis it should be possible to synthesise metabolites such as the α-dinor, α-tetranor, 13-dihydro and 15-oxo derivatives, or combinations of these. The synthesis also lends itself to the production of a variety of analogues. It is thus an extremely attractive proposition for organic chemists who wish to study the relationships between chemical structure and biological activity of the prostaglandins. Details of several syntheses by COREY are contained in the following publications: COREY, 1969; COREY, ANDERSEN, CARLSON, PAUST, VEDEJS, VLATTAS and WINTER, 1968; COREY, VLATTAS, ANDERSEN and HARDING, 1968; COREY, VLATTAS and HARDING, 1969; COREY, WEINSHENKER, SCHAAF and HUBER, 1969; COREY, ARNOLD and HUTTON, 1970; COREY, SCHAAF, HUBER, KOELLIKER and WEINSHENKER, 1970; COREY and NOYORI, 1970).

Other syntheses have been reported but none of these has the attractive features of Corey's. They will not be described in detail here. Syntheses of dl-11-deoxy $PGF_{1\beta}$ (BAGLI, BOGRI, DEGHENGHI and WIESNER, 1966; BAGLI and BOGRI, 1967, 1969), dl-dihydro-PGE_1 ethylester (BEAL, BABCOCK and LINCOLN, 1966), dl-PGB_1 and dl-PGE_1-237 (HARDEGGER, SCHENK and BORGER, 1967; KLOK, PABON and VAN DORP, 1968), dl-PGE_1 and dl-8-iso PGE_1 (SCHNEIDER, AXEN, LINCOLN, PIKE and THOMPSON, 1968, 1969), (15-S)-PGB_1 (COLLINS, JUNG, PAPPO, 1968; PAPPO, COLLINS and JUNG, 1971), dl-13,14-dihydro-PGE_1 (STRIKE and SMITH, 1971) and dl-PGE_1 and dl-$PGF_{1\alpha}$

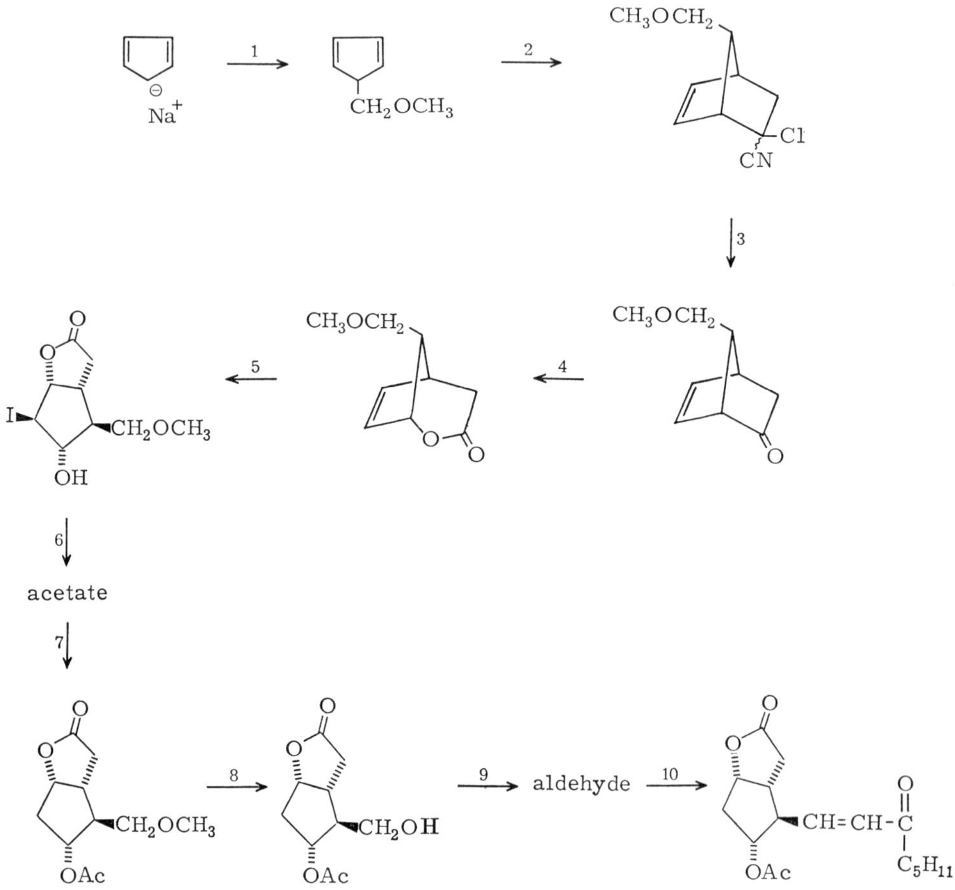

Fig. 1. Synthesis of the primary prostaglandins (E. J. COREY)

1. Slight excess of chloromethyl methyl ether in tetrahydrofuran at 55°, evaporate below 0°
2. Diels-Alder reaction with 2-chloro-acrylonitrile (5 equiv) at 0° in presence of cupric fluoroborate as catalyst
3. Potassium hydroxide (2.5 equiv), hot saturated aqueous solution, in dimethyl sulphoxide for 14 hr at 25—30°
4. 1.25 equiv of m-chloroperbenzoic acid in methylene chloride in the presence of sodium bicarbonate (Bayer-Villager oxidation)
5. Saponification in water containing 2.5 equiv of sodium hydroxide at 0° followed by neutralisation with carbon dioxide and treatment with 2.5 equiv of aqueous potassium tri-iodide solution at 0—5° for 12 hr
6. Acetic anhydride — pyridine at 25° for 15 min

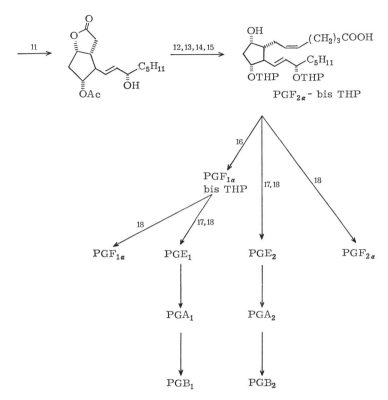

7. Tributyltin hydride in benzene at 25° (initiation with azobisisobutyronitrile)
8. Boron tribromide (5.5 equiv) in methylene chloride at 0°
9. Collins reagent in methylene chloride at 0°
10. Sodio derivative of dimethyl 2-oxoheptyl phosphonate in dimethoxyethane at 25° for 60 min
11. Excess zinc borohydride in dimethoxyethane at 20° for 30 min
12. De-acetylation with equimolar amount of potassium carbonate in methanol at 25° for 15 min
13. Dihydropyran (10 equiv) in methylene chloride containing p-toluenesulphonic acid (0.01 equiv) at 25° for 15 min
14. Di-isobutylaluminium hydride (2 equiv) in toluene at —60° for 30 min
15. Condensed with Wittig reagent derived from 5-triphenylphosphoniopentanoic acid and sodio methylsulphinylcarbamide in dimethylsulphoxide
16. Hydrogenation (1 atm., —15°) in presence of palladium on carbon catalyst, followed by thin layer chromatography.
17. Oxidation with Jones chromic acid reagent.
18. Hydrolysis with acetic acid in water at 39°.

(JUST and SIMONOVITCH, 1967) have been reported. The synthesis by JUST and SIMONOVITCH could not be reproduced by HOLDEN, HWANG, WILLIAMS, WEINSTOCK, HARMAN and WEISBACH (1968) though under modified conditions this method does yield dl-PGF$_{1\alpha}$ and dl-PGF$_{1\beta}$ as their methyl esters (SCHNEIDER et al., 1968; JUST, SIMONOVITCH, LINCOLN, SCHNEIDER, AXEN, SPERO and PIKE, 1969).

Autoxidation of di-homo-γ-linolenic acid yields PGE$_1$ in very small amounts (0.1%) (NUGTEREN, VONKEMAN and VAN DORP, 1967). This reaction is of little practical importance, though the possibility

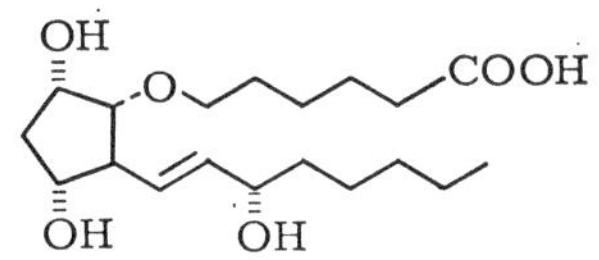

Fig. 2. 7-oxa-prostaglandin F$_{1\alpha}$

that it may result in artifacts must be remembered when prostaglandins are being extracted from tissues rich in unsaturated fatty acid precursors.

Finally, a method has been described (BUNDY, LINCOLN, NELSON, PIKE and SCHNEIDER, 1971) for the production of mammalian-type prostaglandins from 15-epi-PGA$_2$ which can be obtained naturally in large amounts from a gorgonian, *Plexaura homomalla* (WEINHEIMER and SPRAGGINS, 1969). This method has few steps and also has the advantage that very large amounts of starting material can be obtained from natural sources; *Plexaura* contains 1.5% of 15-epi-PGA$_2$ by dry weight.

FRIED has synthesised a series of 7-oxa-prostaglandins (Fig. 2) and some of these have potential importance as prostaglandin antagonists (FRIED, HEIM, SUNDER-PLASSMAN, ETHEREDGE, SANTHANAKRISHNAN and HIMIZU, 1968; FRIED, SANTHANAKRISHNAN, HIMIZU, LIN, FORD, RUBIN and GRIGAS, 1969).

It is clear that within the next decade thousands of prostaglandin analogues will be synthesised. One can therefore anticipate that compounds with more specific and more potent biological activities, as well as prostaglandin antagonists, will become available for both clinical and research use.

References

BAGLI, J. F., BOGRI, T.: Prostaglandins II — An improved synthesis and structural proof of (±)-11-deoxyprostaglandin $F_{1\beta}$. Tetrahedron Lett. No. 1, 5—10 (1967).

— — Prostaglandins III (±)-11-deoxy-13,14-dihydroprostaglandin $F_{1\alpha}$ and $F_{1\beta}$, a novel synthesis of prostanoic acids. Tetrahedron Lett. No. 21, 1639—1646 (1969).

— — DEGHENGHI, R., WIESNER, K.: Prostaglandins I — Total synthesis of 9β, 15-dihydroxyprost-13-enoic acid. Tetrahedron Lett. No. 5, 465—470 (1966).

BEAL, P. F., BABCOCK, J. C., LINCOLN, F. H.: Synthetic approaches in the prostanoic acid series. Nobel Symposium 2, Prostaglandins. Eds.: S. BERGSTRÖM and B. SAMUELSSON. Stockholm: Almqvist and Wiksell 1967, pp. 219—230.

BUNDY, G., LINCOLN, F., NELSON, N., PIKE, T., SCHNEIDER, W.: Novel prostaglandin syntheses. Ann. N. Y. Acad. Sci. 180, 76—90 (1971).

COLLINS, P., JUNG, C. J., PAPPO, R.: Prostaglandin studies. The total synthesis of dl-prostaglandin B_1. Israel J. Chem. 6, 839—841 (1968).

COREY, E. J.: Total syntheses of prostaglandins. In: Proceedings of the Robert A. Welch Foundation Conferences on Chemical Research. XII. Organic Synthesis. Ed.: W. O. MILLIGAN. Houston, Texas, 1969, pp. 51—79.

— ANDERSEN, N. H., CARLSON, R. M., PAUST, J., VEDEJS, E., VLATTAS, I., WINTER, R. E. K.: Total synthesis of prostaglandins. Synthesis of the pure dl-E_1, -$F_{1\alpha}$, -$F_{1\beta}$, -A_1, and -B_1 hormones. J. Amer. chem. Soc. 90, 3245—3247 (1968).

— ARNOLD, Z., HUTTON, J.: Total synthesis of prostaglandins E_2 and $F_{2\alpha}$ (dl) via a tricarbocyclic intermediate. Tetrahedron Lett. No. 4, 307—310 (1970).

— NOYORI, R.: A total synthesis of prostaglandin $F_{2\alpha}$ (dl) from 2-oxabicyclo (3.3.0) oct-6-en-3-one. Tetrahedron Lett. No. 4, 311—313 (1970).

— — SCHAAF, T. K.: Total synthesis of prostaglandins $F_{1\alpha}$, E_1, $F_{2\alpha}$ and E_2 (natural forms) from a common synthetic intermediate. J. Am. chem. Soc. 92, 2586—2587 (1970).

— VLATTAS, I., ANDERSEN, N. H., HARDING, K.: A new total synthesis of prostaglandins of the E_1 and F_1 series including 11-epiprostaglandins. J. Am. chem. Soc. 90, 3247—3248 (1968) (see erratum 90, 5947 [1968]).

— — HARDING, K.: Total synthesis of natural (levo) and enantiomeric (dextro) forms of prostaglandin E_1. J. Am. chem. Soc. 91, 535—536 (1969).

— SCHAAF, T. K., HUBER, W., KOELLIKER, U., WEINSHENKER, N. M.: Total synthesis of prostaglandins $F_{2\alpha}$ and E_2 as the naturally occurring forms. J. Am. chem. Soc. 92, 397—398 (1970).

— WEINSHENKER, N. M., SCHAAF, T. K., HUBER, W.: Stereo-controlled synthesis of prostaglandins $F_{2\alpha}$ and E_2 (*dl*). J. Am. chem. Soc. **91**, 5675—5677 (1969).

FRIED, J., HEIM, S., SUNDER-PLASSMAN, P., ETHEREDGE, S. J., SANTHANA-KRISHNAN, T. S., HIMIZU, J.: Synthesis af 15-desoxy-7-oxa-prostaglandin $F_{1\alpha}$ and related substances. Prostaglandin Symposium of the Worcester Foundation for Exp. Biol. Eds.: P. W. RAMWELL and J. E. SHAW. New York: Interscience 1968, pp. 351—363.

— SANTHANAKRISHNAN, T. S., HIMIZU, J., LIN, C. H., FORD, S. H., RUBIN, B., GRIGAS, E. O.: Prostaglandin antagonists: synthesis and smooth muscle activity. Nature. **223**, 208—210 (1969).

HARDEGGER, E., SCHENK, H. P., BROGER, E.: Synthese der DL-Form eines natürlichen Prostaglandins. Helv. chim. Acta **50**, 2501—2504 (1967).

HOLDEN, K. G., HWANG, B., WILLIAMS, K. R., WEINSTOCK, J., HARMAN, M., WEISBACH, J. A.: Synthetic studies on prostaglandins. Tetrahedron Lett. No. **13**, 1569—1574 (1968).

JUST, G., SIMONOVITCH, C.: A prostaglandin synthesis. Tetrahedron Lett. No. **22**, 2093—2097 (1967).

— — LINCOLN, F. H., SCHNEIDER, W. P., AXEN, U., SPERO, G. B., PIKE, J. E.: A synthesis of prostaglandin $F_{1\alpha}$ and related substances. J. Am. chem. Soc. **91**, 5364—5371 (1969).

KLOK, R., PABON, H. J. J., VAN DORP, D. A.: Synthesis of DL-prostaglandin B_1 and its reduction product DL-prostaglandin E_1-237. Recl. Trav. chim. Pays-Bas Belg. **87**, 813—823 (1968).

NUGTEREN, D. H., VONKEMAN, H., VAN DORP, D. A.: Non-enzymic conversion of all-*cis*-8,11,14-eicosatrienoic acid into prostaglandin E_1. Recl. Trav. chim. Pays-Bas Belg. **86**, 1237—1245 (1967).

PAPPO, R., COLLINS, P. W., JUNG, C. J.: New synthetic approach in the prostaglandin field. Ann. N. Y. Acad. Sci. **180**, 64—75 (1971).

SCHNEIDER, W. P., AXEN, U., LINCOLN, F. H., PIKE, J. E., THOMPSON, J. L.: The total synthesis of prostaglandins. J. Am. chem. Soc. **90**, 5895—5896 (see erratum **91**, 1043) (1968).

— — — — — The synthesis of prostaglandin E_1 and related substances. J. Am. chem. Soc. **91**, 5372—5378 (1969).

STRIKE, D. P., SMITH, H.: A novel total synthesis of $\pm$-13,14-dihydro prostaglandin. Ann. N. Y. Acad. Sci. **180**, 91—100 (1971).

WEINHEIMER, A. J., SPRAGGINS, R. L.: The occurrence of two new prostaglandin derivatives (15-epi-PGA_2 and its acetate, methyl ester) in the gorgonian *Plexaura homomalla*. Tetrahedron Lett. No. **59**, 5185—5188 (1969).

IV. Biosynthesis

The enzymatic conversion of all-*cis*-5,8,11,14-eicosatetraenoic acid (arachidonic acid) to PGE_2 was discovered independently by D. A. van Dorp in the Netherlands and S. Bergström in Sweden (van Dorp, Beerthuis, Nugteren and Vonkeman, 1964 a; Bergström, Danielsson and Samuelsson, 1964). $PGF_{2\alpha}$ is also formed from arachidonic acid (Änggård and Samuelsson, 1965) whilst PGE_1

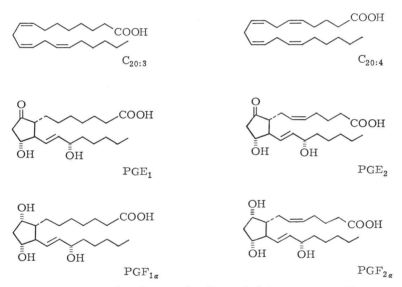

Fig. 1. Formulae of prostaglandins and their precursor acids

and $PGF_{1\alpha}$ are similarly formed from 8,11,14-eicosatrienoic acid (dihomo-γ-linolenic acid), the immediate precursor of arachidonic acid (van Dorp, Beerthuis, Nugteren and Vonkeman, 1964 b; Bergström et al., 1964; Kupiecki, 1965) (Fig. 1). The conversion of all-*cis*-5,8,11,14,17-eicosapentaenoic acid to PGE_3 has also been demonstrated (Bergström et al., 1964).

The biosynthetic pathway for PGA_1 and PGA_2 has not yet been elucidated although the formation of PGA_1 has been reported during incubation of 8,11,14-eicosatrienoic acid with sheep seminal vesicles (DANIELS, HINMAN, JOHNSON, KUPIECKI, NELSON and PIKE, 1965). The facility with which these compounds can be derived chemically from PGE_1 and PGE_2 suggests the possibility that PGA compounds are formed by an enzymatically-catalysed dehydration of PGE compounds.

An enzyme which converts PGA_1 and PGA_2 to PGB_1 and PGB_2 (Fig. 2) has been discovered in plasma by R. L. JONES (JONES, 1970). This enzyme has so far been found in the plasma of dog, pig, cat, rabbit but not guinea-pig or man. Its properties are discussed more fully in Chapter V.

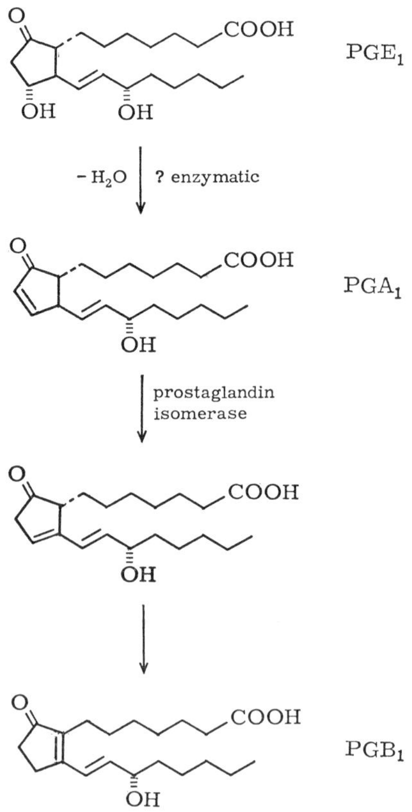

Fig. 2. Pathway postulated for bioconversion of PGE_1 to PGA_1 and PGB_1

1. Prostaglandin Synthetase

The enzyme system which catalyses the conversion of unsaturated fatty acids to prostaglandins has been named prostaglandin synthetase. This enzyme is found in many tissues (Table 1) but a particularly rich source is the sheep vesicular gland, homogenates of which have been used extensively for the large-scale production of prostaglandins (see below).

Table 1. *Tissues shown to contain prostaglandin synthetase*

Prostate	human [1]
Seminal vesicles	sheep, ox [2], human [1]
Lung	guinea-pig, sheep
Intestine	sheep, guinea-pig
Uterus	human (endometrium), sheep, guinea-pig
Thymus	sheep
Heart	sheep
Liver	sheep, guinea-pig, rat
Kidney	sheep, guinea-pig, rat
Pancreas	sheep
Brain	guinea-pig
Stomach	rat [3]
Iris	pig [4]

Data from VAN DORP (1966) except as indicated by references below:
[1] VAN DORP (1967).
[2] WALLACH (1965).
[3] PACE-ASCIAK, MORAWSKA, COCEANI and WOLFE (1968).
[4] VAN DORP, JOUVENAZ and STRUIJK (1967).

Prostaglandin synthetase is associated with the microsomal fraction, but so far no reports have been published of preparations with high purity. An enzyme-rich particulate fraction has been obtained by centrifugation of sheep vesicular gland homogenates for 1 hour at 100,000 g after removal of other cell particles at lower speeds. The freeze-dried sediment retains its activity for several months when stored at $-20°$.

The addition of glutathione greatly enhances the yield of PGE compounds though this is partly at the expense of the corresponding PGF. Other SH-containing compounds (cysteine, homo-cysteine, thiophenol, thioglycollic acid) are far less effective than glutathione.

Hydroquinone also increased the yield moderately but neither ATP nor NADH had any effect. Although less effective than glutathione, ascorbic acid and various phenols all increase the yield of prostaglandin (van Dorp, 1966, 1967; Nugteren, Beerthuis and van Dorp, 1966). The enzyme is partially inhibited by divalent cations such as copper, zinc and cadmium (Nugteren et al., 1966).

The formation of prostaglandins in low yield from precursor acids in the presence of oxygen can occur in the absence of an enzyme (Nugteren, Vonkeman and van Dorp, 1967).

2. Formation of Products Other than PGE or PGF

On incubation of all-cis-8,11,14-eicosatrienoic acid with sheep vesicular gland homogenates five products have been reported in

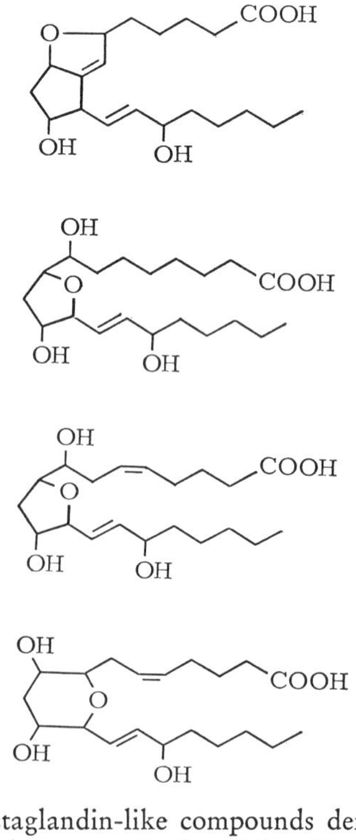

Fig. 3. Additional prostaglandin-like compounds derived from arachidonic acid on incubation with ram seminal vesicles (Pace-Asciak and Wolfe, 1971)

addition to PGE$_1$ and PGF$_{1\alpha}$. These include 11-dehydro-F$_{1\alpha}$ (conclusively identified by GRANSTRÖM, LANDS and SAMUELSSON, 1968), 11α-hydroxy-9,15-di-oxo-13-*trans*-prostenoic acid, 11-hydroxy-8-*cis*-12-*trans*-14-*cis*-eicosatrienoic acid, 12-hydroxy-8-*trans*-10-*trans*-hep-

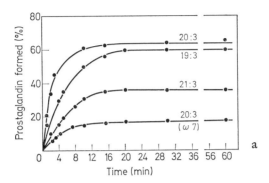

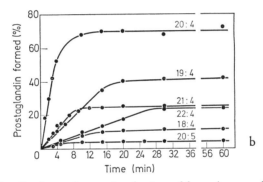

Fig. 4. Prostaglandin formation on incubation of homologues of di-homo-γ-linolenic acid (a) and of arachidonic acid (b) with sheep seminal vesicle homogenate. Ordinate: % prostaglandin formed. Abccissa: time in minutes (Data from STRUIJK et al., 1967)

tadecadienoic acid and 15-hydroperoxy-11α-hydroxy-9-oxo-13-*trans*-prostenoic acid (NUGTEREN, BEERTHUIS and VAN DORP, 1966).

PACE-ASCIAK and WOLFE (1971) have recently reported four new metabolites of arachidonic acid found on incubation with the sheep vesicular gland enzyme system. The proposed structures of these are shown in Fig. 3.

3. Substrate Specificity

The structural requirements for the biosynthesis of prostaglandin-type compounds from unsaturated straight chain acids has been studied using a variety of precursors. Of the all-*cis*-tetraenoic acids tested, the C20 and C19 homologues gave high yields of prostaglandin, PGE_2 (71%) and α-nor-PGE_2 (41%) respectively, the C21 and C22 gave somewhat lower yields (about 25%). The time course of the formation is shown in Fig. 4 and the yields are given in Table 2. Neither the esters nor the alcohols (all-*cis*-8,11,14-eicosatrienol and

Table 2. *Yields of prostaglandins on incubating sheep vesicular gland homogenate with different substrates*

ω 6 Fatty Acids	PGE formed %	Δ 3 Fatty Acids	PGE formed %
18 : 3	< 5	18 : 3 ω 4	0
19 : 3	60	18 : 3 ω 6	< 5
20 : 3	68	19 : 3 ω 5	0
21 : 3	36	19 : 3 ω 6	60
22 : 3	< 5	20 : 3 ω 5	< 2
18 : 4	11	20 : 3 ω 6	68
19 : 4	41	20 : 3 ω 7	18
20 : 4	71	20 : 3 ω 8	0
21 : 4	25	20 : 3 ω 9	0
22 : 4	24	22 : 3 ω 6	0

Table 3. *The effect on PGE formation of introducing a trans double bond into the precursor acid* (VAN DORP, 1971)

Precursor Fatty Acid (C20)	EFA Activity (18 : 2 ω 6 = 100)	PGE Formed (%) (20 : 3 ω 6 = 100)	PG Formed
8 c, 11 c, 14 c	102	100	PGE_1
2 t, 8 c, 11 c, 14 c		16	2 t PGE
3 t, 8 c, 11 c, 14 c		66	3 t PGE
4 c, 8 c, 11 c, 14 c	67	66	4 c PGE
5 c, 8 c, 11 c, 14 c	106	107	PGE_2
5 t, 8 c, 11 c, 14 c		46	5 t PGE
8 c, 11 c, 14 c, 18 c	7	6	18 c PGE_1

all-*cis*-5,8,11,14-eicosatetraenol) are substrates for the enzyme (STRUIJK, BEERTHUIS and VAN DORP, 1967). Other substrates for prostaglandin synthetase are shown in Table 3.

4. Inhibitors of Prostaglandin Synthetase

Certain analogues of the essential fatty acids act as inhibitors of prostaglandin synthetase. In particular 8-*cis*-12-*trans*-14-*cis*-eicosatrienoic acid and 5-*cis*-8-*cis*-12-*trans*-14-*cis*-eicosatetraenoic acid are effective blocking agents *in vitro*. The former of these has been fed to rats at 70 mg/day; growth rate was reduced by 20%, but little of the abnormal acid was incorporated into the lipids of the liver (VAN DORP, 1971; NUGTEREN, 1970).

Linoleic (18 : 2 ω 6) and α-linolenic (18 : 3 ω 3) acids at a concentration of 3.6 mM inhibit the action of prostaglandin synthetase on arachidonic acid. This has been shown with ram seminal vesicles and with rat stomach homogenates (PACE-ASCIAK and WOLFE, 1968). Oleic acid was less active as an enzyme inhibitor.

Recently the acetylenic anologue of arachidonic acid, 5,8,11,14-eicosatetraynoic acid, has been shown to inhibit prostaglandin biosynthesis (AHERN and DOWNING, 1970). Furthermore anti-inflammatory drugs like aspirin and indomethacin also block prostaglandin synthesis (SMITH and WILLIS, 1971; VANE, 1971).

5. Origin of Substrate

There is relatively little free arachidonic acid and still less dihomo-γ-linolenic acid in cells. These acids are incorporated into membrane phospholipids from which they must be hydrolysed before prostaglandin synthetase can act. It has never been possible to detect any prostaglandin in the phospholipid fraction and none of the newly synthesised prostaglandin is incorporated into the phospholipid (LANDS and SAMUELSSON, 1968; VONKEMAN and VAN DORP, 1968). BARTELS, VOGT and WILLE (1968) have shown that phospholipase A injection will increase prostaglandin output, presumably by making the substrate available for prostaglandin synthetase. ELIASSON (1958) made similar observations *in vitro*. It is probable that activation of phospholipase A may be the rate limiting step in prostaglandin synthesis (KUNZE and BOHN, 1969).

6. Mechanism of Prostaglandin Formation

In 1962 VAN DORP made the following comments about the conformation of arachidonic acid in the liquid state; "It appeared that there is considerable steric hindrance if one tries to rotate one of the legs of the U caused by this peculiar combination of *cis*-double bonds and methylene groups. This may mean that once the U form is present, it will stay in this shape because there are energy barriers to be overcome to get out of this conformation. We are inclined to believe that this may have biochemical significance. For instance, is this the reason why the double bonds should be at the terminal 6 and 9 position? Only then would a sufficiently bent conformation occur which makes dehydrogenations at the terminal 12 and 15 places possible." With this in mind when he read the Swedish work on the structure of prostaglandins, VAN DORP reasoned that essential fatty acids such as arachidonic acid might act as precursors of prostaglandins by oxidative ring closure (VAN DORP, 1966).

By using 8,11,14-eicosatrienoic acid tritiated at either the 8,11 or 12 position, KLENBERG and SAMUELSSON (1965) demonstrated that the hydrogens at these positions in the precursor acid are retained in the same positions in the PGE_1 molecule following bond formation between the carbon atoms at the 8 and 12 positions.

Using $^{18}O_2$ gas in the biosynthesis of PGE_1 it has been conclusively proved by mass spectrometric analysis that the oxygen atoms at the 11 and 15 positions in PGE_1 are derived from molecular oxygen (NUGTEREN and VAN DORP, 1965; RYHAGE and SAMUELSSON, 1965). SAMUELSSON (1965) repeated the experiment using oxygen gas with the following composition ($^{16}O^{16}O$ 43%; $^{16}O^{18}O$ 1%; $^{18}O^{18}O$ 56%) and prevented exchange of the ketonic oxygen formed during incubation by immediate reduction with $NaBH_4$. He thus established that the oxygen at C 9 is derived from the same molecule of oxygen as the oxygen atom in the C 11 hydroxyl. These important results were confirmed by NUGTEREN, BEERTHUIS and VAN DORP (1966) using a slightly different procedure. HAMBERG and SAMUELSSON (1967 a)

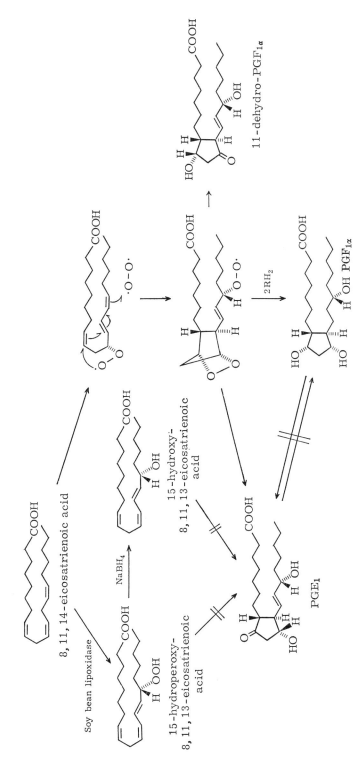

Fig. 5. Pathways in bioconversion of dihomo-γ-linolenic acid to PGE₁ and PGF₁α by sheep vesicular gland homogenate

used 13D and 13L tritiated 8,11,14-eicosatrienoic acid to show that
the hydrogen lost from C 13 during the biosynthesis of PGE_1 has the
L configuration. Neither 15L-hydroperoxy-8-*cis*-11-*cis*-13-*trans*-eico-
satrienoic acid nor 15-L-hydroxy-8-*cis*-11-*cis*-13-*trans*-eicosatrienoic
acid was converted to PGE_1 or $PGF_{1\alpha}$ (Fig. 5). Furthermore using
precursor acid tritiated in the 9 position, it was proved that $PGF_{1\alpha}$
is not formed via PGE_1.

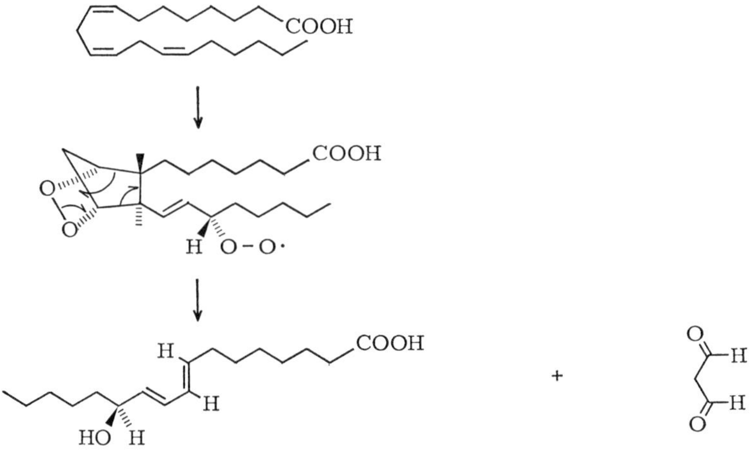

Fig. 6. Formation of 12-hydroxy-8,10-heptadecadienoic acid and malonaldehyde from the endoperoxide of 8,11,14-eicosatrienoic acid

It is proposed that the first intermediary in the biosynthesis is
11-peroxy-8,12,14-eicosatrienoic acid and that this is cyclized into an
endoperoxide. This second intermediary is then transformed by inde-
pendent pathways to PGE_1 and $PGF_{1\alpha}$ (Fig. 5).

Further evidence in favour of this mechanism comes from the
observation, mentioned above, that oxygenation of 8,11,14-eicosatri-
enoic acid in the presence of the sheep vesicular gland enzyme yields
several products in addition to PGE_1 and $PGF_{1\alpha}$. One of these is
12-hydroxy-8,10-heptadecadienoic acid (Fig. 6). By means of 8,11,14-
eicosatrienoic acids tritiated at either 9,10,11,13 (D- or L-) or 15 positions, it was shown that carbon atoms 9, 10 and 11 of the precursor
are eliminated as malonaldehyde and that the hydrogen removal from
C 13 is stereochemically identical with that occurring during the bio-

synthesis of PGE_1. It seems very probable that both prostaglandins and 12-hydroxy-8,10-heptadecadienoic acid originate from the same cyclic peroxide of 8,11,14-eicosatrienoic acid (HAMBERG and SAMUELSSON, 1966, 1967 b).

7. Large Scale Biosynthetic Production of Prostaglandins

Prostaglandins for biological, clinical or metabolic investigations have been prepared on large scale by several commerical laboratories. The method used by the Upjohn Company for the preparation of PGE_2 (DANIELS and PIKE, 1968) will be described briefly here (see Fig. 7).

Frozen sheep vesicular gland tissue (75 kg) is ground and homogenised in 0.1 M NH_4Cl containing 500 mg of reduced glutathione, 50 mg of hydroquinone and 10 ml of 0.1 M disodium E.D.T.A./litre of buffer. The homogenate is incubated with 37 g of arachidonic acid with continuous aeration at 37° for 1 hour, after adjusting the pH to 8.5 with NH_4OH.

At the end of the incubation 3 volumes of cold acetone are added, the mixture being stirred for a further hour before filtration. The filtrate is concentrated to one fifth of its original volume, so that the residue is predominantly aqueous. This is adjusted to pH 7.0 and extracted with Skerrysolve B. The aqueous phase is then adjusted to pH 3.0 with citric acid and extracted with methylene chloride.

Approximately 60 g of crude lipid is obtained, containing 8 to 12 g of a prostaglandin mixture. The yield on such large scale preparations varies between 15 and 30%. Further purification was achieved by silica gel column chromatography (10 g lipid/100 g silica gel), the column being eluted with increasing concentrations of methanol in chloroform. The PGE compounds are eluted with 4% MeOH and the PGF compounds with 10% MeOH in $CHCl_3$. Since the sheep vesicular gland homogenate contain pre-formed PGE_1 and PGF compounds these must be separated from the PGE_2.

Separation of PGE_1 and PGE_2 (or $PGF_{1\alpha}$ and $PGF_{2\alpha}$) eluted from the silica gel was achieved on the gram scale by using a column of Amberlyst 15 resin. The resin is first completely converted to the silver cycle with 5% aqueous silver nitrate and washed with ethanol. Up to 2 g prostaglandin mixture can be loaded on a 100 g column.

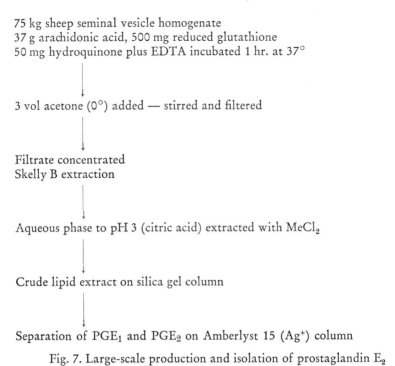

75 kg sheep seminal vesicle homogenate
37 g arachidonic acid, 500 mg reduced glutathione
50 mg hydroquinone plus EDTA incubated 1 hr. at 37°

↓

3 vol acetone (0°) added — stirred and filtered

↓

Filtrate concentrated
Skelly B extraction

↓

Aqueous phase to pH 3 (citric acid) extracted with $MeCl_2$

↓

Crude lipid extract on silica gel column

↓

Separation of PGE_1 and PGE_2 on Amberlyst 15 (Ag^+) column

Fig. 7. Large-scale production and isolation of prostaglandin E_2

PGE_1 is eluted with ethanol and then PGE_2 is eluted with ethanol containing 5% cyclohexene. Prostaglandins isolated by this procedure are re-crystallized from an ethyl acetate-pentane mixture.

Other prostaglandins or their derivatives can be formed chemically from these biosynthetic products. Thus reduction of PGE_2 with sodium borohydride yields a mixture of $PGF_{2\alpha}$ (m.p. 35—37°) and $PGF_{2\beta}$ (m.p. 93—94°). These may be separated by reversed phase chromatography or by silica gel if ethyl acetate-benzene mixtures are used. PGA_2 is readily formed by acetic acid dehydration of PGE_2.

References

AHERN, D. G., DOWNING, D. T.: Inhibition of prostaglandin biosynthesis in sheep seminal vesicular tissue by eicosa-5:8:11:14-tetraynoic acid. Fedn Proc. Fedn Am. Socs exp. Biol. **29**, 854 Abs. (1970).

ÄNGGÅRD, E., SAMUELSSON, B.: Biosynthesis of prostaglandins from arachidonic acid in guinea-pig lung. J. biol. Chem. **240**, 3518—3521 (1965).

BARTELS, J., VOGT, W., WILLE, G.: Prostaglandin release from and formation in perfused frog intestine. Arch. Pharmak. exp. Path. **259**, 153—154 (1968) [see erratum **259**, 459 (1968)].

BERGSTRÖM, S., DANIELSSON, H., SAMUELSSON, B.: The enzymatic formation of prostaglandin E$_2$ from arachidonic acid. Biochim. biophys. Acta **90**, 207—210 (1964).

DANIELS, E. G., HINMAN, J. W., JOHNSON, B. A., KUPIECKI, F. P., NELSON, J. W., PIKE, J. E.: The isolation of an additional prostaglandin derivative from the enzymic cyclization of homo-γ-linolenic acid. Biochem. biophys. Res. Comm. **21**, 413—417 (1965).

— PIKE, J. E.: Isolation of prostaglandins. Prostaglandin Symposium of the Worcester Foundation for Exp. Biol. Ed.: P. W. RAMWELL and J. E. SHAW. New York: Interscience 1968, pp. 379—387.

ELIASSON, R.: Formation of prostaglandin *in vitro*. Nature **182**, 256—257 (1958).

GRANSTRÖM, E., LANDS, W. E. M., SAMUELSSON, B.: Biosynthesis of 9α,15-dihydroxy-11-keto-prost-13-enoic acid. J. biol. Chem. **243**, 4104—4108 (1968).

HAMBERG, M., SAMUELSSON, B.: Novel biological transformations of 8,11,14-eicosatrienoic acid. J. Am. chem. Soc. **88**, 2349—2350 (1966).

— — On the mechanism of the biosynthesis of prostaglandins E$_1$ and F$_{1\alpha}$. J. biol. Chem. **242**, 5336—5343 (1967 a).

— — Oxygenation of unsaturated fatty acids by the vesicular gland of sheep. J. biol. Chem. **242**, 5344—5354 (1967 b).

JONES, R. L.: A prostaglandin isomerase in cat plasma. Biochem. J. **119**, 64—65 P (1970).

KLENBERG, D., SAMUELSSON, B.: The biosynthesis of prostaglandin E$_1$ studied with specifically ^{3}H-labelled 8,11,14-eicosatrienoic acids. Acta chem. scand. **19**, 534—535 (1965).

KUNZE, H., BOHN, R.: Formation of prostaglandin in bovine seminal vesicles. Arch. Pharmak. exp. Path. **264**, 263—264 (1969).

KUPIECKI, F. P.: Conversion of homo-γ-linolenic acid, to prostaglandin F$_{1\alpha}$ by ovine and bovine seminal vesicle extracts. Life Sci. **4**, 1811—1815 (1965).

LANDS, W. E. M., SAMUELSSON, B.: Phospholipid precursors of prostaglandins. Biochim. biophys. Acta **164**, 426—429 (1968).

NUGTEREN, D. H.: Inhibition of prostaglandin biosynthesis by 8*cis*, 12*trans*,14*cis*-eicosatrienoic acid and 5*cis*, 8*cis*,12*trans*, 14*cis*-eicosatetraenoic acid. Biochim. biophys. Acta **210**, 171—176 (1970).

— BEERTHUIS, R. K., VAN DORP, D. A.: The enzymic conversion of all-*cis*-8,11,14-eicosatrienoic acid into prostaglandin E$_1$. Recl. Trav. chim. Pays-Bas Belg. **85**, 405—419 (1966).

— VAN DORP, D. A.: The participation of molecular oxygen in the biosynthesis of prostaglandins. Biochim. biophys. Acta **98**, 654—656 (1965).

NUGTEREN, D. H., VONKEMAN, H., VAN DORP, D. A.: Non-enzymic conversion of all-*cis*-8,11,14-eicosatrienoic acid into prostaglandin E_1. Recl. Trav. chim. Pays-Bas Belg. **86**, 1237—1245 (1967).

PACE-ASCIAK, C., MORAWSKA, K., COCEANI, F., WOLFE, L. S.: The biosynthesis of prostaglandins E_2 and F_{2a} in homogenates of the rat stomach. Prostaglandin Symposium of the Worcester Foundation for Exp. Biol. Ed.: P. W. RAMWELL and J. E. SHAW. New York: Interscience 1968, pp. 371—378.

— WOLFE, L. S.: Inhibition of prostaglandin synthesis by oleic, linoleic and linolenic acids. Biochim. biophys. Acta **152**, 784—787 (1968).

— — Ann. N. Y. Acad. Sci. **180**, 199 (1971).

RYHAGE, R., SAMUELSSON, B.: The origin of oxygen incorporated during the biosynthesis of prostaglandin E_1. Biochem. biophys. Res. Comm. **19**, 279—282 (1965).

SAMUELSSON, B.: On the incorporation of oxygen in the conversion of 8,11,14-eicosatrienoic acid to prostaglandin E_1. J. Am. chem. Soc. 87, 3011—3013 (1965).

SMITH, J. B., WILLIS, A. L.: Aspirin selectively inhibits prostaglandin production in human platelets. Nature **231**, 235—237 (1971).

STRUIJK, C. B., BEERTHUIS, R. K., VAN DORP, D. A.: Specificity in the enzymatic conversion of poly-unsaturated fatty acids into prostaglandins. Nobel Symposium 2, Prostaglandins. Ed.: S. BERGSTRÖM and B. SAMUELSSON. Stockholm: Almqvist and Wiksell 1967, pp. 51—56.

VAN DORP, D. A.: Properties and synthesis of certain biologically important unsaturated fatty acids. VIth I. S. F. Congress, London. 9—13 April (1962).

— The biosynthesis of prostaglandins. Mem. Soc. Endocr. **14**, 39—47 (1966).

— Aspects of the biosynthesis of prostaglandins. Prog. biochem. Pharmac. **3**, 71—82 (1967).

— Recent developments in the biosynthesis and the analyses of prostaglandins. Ann. N. Y. Acad. Sci. **180**, 181—195 (1971).

— BEERTHUIS, R. K., NUGTEREN, D. H., VONKEMAN, H.: The biosynthesis of prostaglandins. Biochim. biophys. Acta **90**, 204—207 (1964 a).

— — — — Enzymatic conversion of all-*cis*-polyunsaturated fatty acids into prostaglandins. Nature **203**, 839—841 (1964 b).

— JOUVENAZ, G. H., STRUIJK, C. B.: The biosynthesis of prostaglandin in pig eye iris. Biochim. biophys. Acta **137**, 396—399 (1967).

VANE, J. R.: Inhibition of prostaglandin synthesis as a mechanism of action for aspirin-like drugs. Nature **231**, 232—235 (1971).

VONKEMAN, H., VAN DORP, D. A.: The action of prostaglandin synthetase on 2-arachidonyl-lecithin. Biochim. biophys. Acta **164**, 430—432 (1968).

WALLACH, D. P.: The enzymic conversion of arachidonic acid to prostaglandin E_2 with acetone powder preparations of bovine seminal vesicles. Life Sci. **4**, 361—364 (1965).

V. Metabolism and Fate

1. Autoradiographic Studies in Mice

Fifteen minutes after the intravenous injection of $PGF_{2\alpha}$ (tritiated at C-17,18) in mice, an autoradiograph of sagittal sections showed that most of the radioactivity was concentrated in the liver, kidneys and subcutaneous connective tissue (Fig. 1). No significant amounts could be demonstrated in the heart, brain, adipose tissue or endocrine glands (GRÉEN, HANSSON and SAMUELSSON, 1967).

Experiments with PGE_1 (tritiated at the C-5,6) yielded similar results (HANSSON and SAMUELSSON, 1965). As early as 2 minutes after an intravenous injection of PGE_1 most of the radioactivity is present in the liver, subcutaneous connective tissue, kidney and myometrium (the endometrial content being low), a picture which is essentially the same 30 minutes after the injection (Fig. 2). One hour after the injection appreciable excretion of radioactivity into the bile and urine has already occurred; the labelled material is now present in liver, bile (the latter being both in the gall bladder and intestinal lumen), kidneys, urine (in the urinary bladder) and thoracic duct indicating absorption from the intestinal lumen. There is radioactivity in the connective tissue but not the smooth muscle of the intestinal wall. Early after the injection both blood and lungs contain moderate amounts of radioactivity.

There is little or no radioacitivity at any time (from 2 to 60 min after injection) in nervous tissue, blood vessels, myocardium, lymph nodes, thymus, spleen, pituitary, adrenals, thyroid, salivary glands, pancreas (including the islets of Langerhans), adipose tissue, testis, epididymis or seminal vesicle.

The rapid uptake of radioactivity by mouse liver and kidney and its excretion via bile and urine agrees with findings in the rat (SAMUELSSON, 1964). The substantial uptake by loose subcutaneous tissue which stains pink by the van Grisen method merits further study as suggested by SAMUELSSON in 1964.

5*

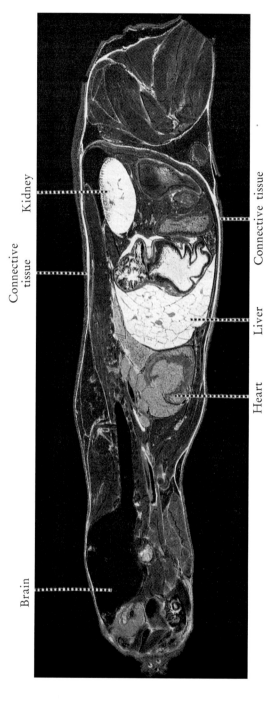

Fig. 1. Autoradiograph showing distribution of radioactivity (light areas) in a male mouse 15 minutes after intravenous injection of ³H-labelled PGF$_{2\alpha}$ (4 μg, 140 μCi/μg). Sagittal section through the entire body of the animal. Note the high concentration of radioactivity in the liver, kidney and connective tissue. (GRÉEN, HANSSON and SAMUELSSON, 1967)

Brain Connective
tissue Kidney

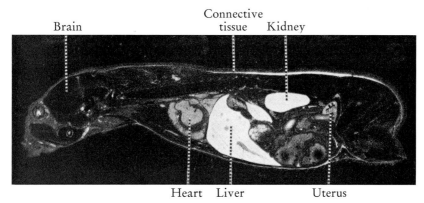

Heart Liver Uterus

Fig. 2. Autoradiograph showing the distribution of radioactivity (light areas) in a female mouse 15 minutes after intravenous injection of ^{3}H-labelled PGE_1 (7 μg). Sagittal section through the entire body of the animal. Note the high concentration of radioactivity in the liver, kidney, subcutaneous connective tissue and uterus. (HANSSON and SAMUELSON, 1965)

2. Distribution and Fate in the Rat

Labelled PGE_1 was infused intravenously for 20 minutes in female rats and the urinary and faecal excretion of radioactive material was followed (SAMUELSSON, 1964). About two-thirds of the administered radioactivity was recovered, of which 85% was excreted in the urine mostly in the first 20 hours. The remainder was recovered from the faeces over a period of 42 hours. When the bile duct was cannulated there was negligible radioactivity in the faeces and about one-fifth of the total radioactivity recovered was present in the bile.

Following subcutaneous injections of PGE_1 in female rats, the concentration of labelled material in different tissues was estimated at different time intervals after the injection. It is remarkable that high levels of radioactivity were already present in the kidney one minute after subcutaneous injection. The levels in the kidney rose rapidly, reaching a peak after 15 minutes and always exceeding the levels found in any other tissue. Radioactivity accumulated in the liver more slowly reaching a peak after approximately 20 minutes. These two organs accounted for most of the injected PGE_1. The lungs contained moderate amounts with a maximum at 10 to 15 minutes after injection. Small amounts accumulated rapidly in

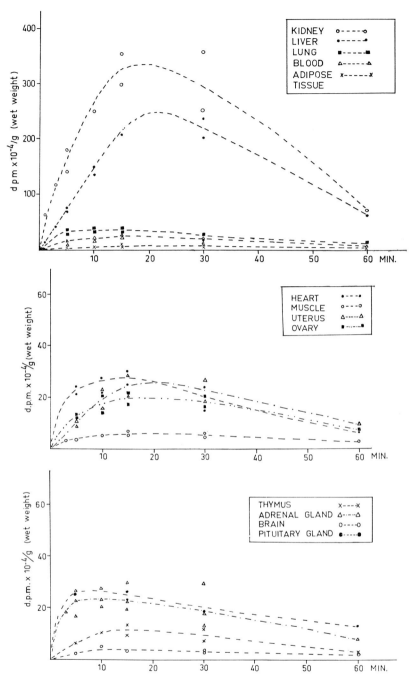

Fig. 3. Distribution of tritium after subcutaneous injection of ³H-PGE₁ (0.2 µg; 140 µCi/µg) to female rats. The rats were killed after the injection at the times indicated. Each point represents means of duplicate determinations of pooled material from four animals. (a) kidney, liver, lung, blood and adipose tissue; (b) heart, skeletal muscle, uterus and ovary; (c) thymus, adrenal, brain and pituitary. (SAMUELSSON, 1964)

the heart, pituitary and adrenals and more slowly in the uterus. Brain, adipose tissue and skeletal muscle contained barely detectable amounts at any time during the experiment. After 60 minutes, any radioactivity not yet excreted was mainly concentrated in the kidney and liver (Fig. 3).

Plasma collected from rats 15 minutes after the PGE_1 injection contained no unchanged labelled PGE_1 but two labelled metabolites were isolated and identified as 11α,15α-dihydroxy-9-oxo-prostanoic acid and 11α-hydroxy-9,15-di-oxo-prostanoic acid which were first demonstrated in homogenates of guinea-pig lung (ÄNGGÅRD and SAMUELSSON, 1964).

Part of the labelled material in liver 15 minutes after the injection was unchanged PGE_1 but the major part like that in the bile was present as more polar compounds. Neither of the metabolites isolated from plasma could be detected in liver. There were also more polar derivatives in the kidney and urine but these have not been identified.

When $^{14}C\text{-}PGE_1$ is injected into the blood-perfused isolated rat liver, metabolites which behave chromatographically like dinor-PGE_1 and tetranor-PGE are formed (DAWSON, JESSUP, McDONALD-GIBSON, RAMWELL and SHAW, 1970).

3. Urinary Metabolites in the Rat

When tritium-labelled $PGF_{1α}$ is injected subcutaneously in rats, about a third of the radioactivity can be recovered from the urine, the major metabolite being α-dinor-$PGF_{1α}$ (GRANSTRÖM, INGER and SAMUELSSON, 1965; GRÉEN and SAMUELSSON, 1968). The same urinary metabolite is excreted following $PGF_{2α}$ administration, but other metabolites including α-tetranor-PGF are also excreted (Fig. 4) (GRÉEN, 1969). Numerous metabolites of PGE_2 have been identified in rat urine (Fig. 5) (SAMUELSSON, GRANSTRÖM, GRÉEN and HAMBERG, 1971).

4. Metabolism in the Guinea-pig

Enzymes in the supernatant fraction of guinea-pig lung homogenates convert PGE_1 to two metabolites, 11α,15α-dihydroxy-9-oxo-prostanoic acid and 11α-hydroxy-9,15-di-oxo-prostanoic acid (ÄNG-

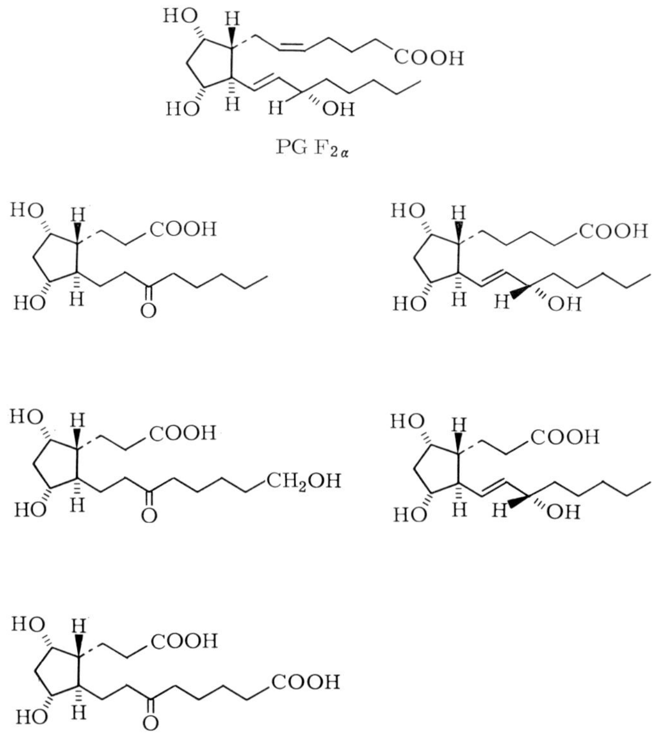

Fig. 4. Rat urinary metabolites of PGF$_{2\alpha}$

GÅRD and SAMUELSSON, 1964). PGE$_2$ and PGE$_3$ yield analogous
metabolites (ÄNGGÅRD, GRÉEN and SAMUELSSON, 1965; ÄNGGÅRD
and SAMUELSSON, 1965) (Fig. 6).

The principal urinary metabolites of PGE$_2$ and PGF$_{2\alpha}$ in the
guinea-pig are 5β,7α-dihydroxy-11-oxo-tetranor-prostanoic acid and
5α,7α-dihydroxy-11-oxo-tetranor-prostanoic acid respectively (GRAN-
STRÖM and SAMUELSSON, 1969 a; HAMBERG and SAMUELSSON, 1969 a;
SAMUELSSON et al., 1971). Oxidation of the secondary alcohol at the
C-15 position and reduction of the *trans*-double bond occur readily
in the lungs, β-oxidation of the carboxyl side chain probably occurs
in the liver (by analogy with the rat).

Reduction of the oxo group in the five membered ring has not
been encountered so far in other species. It is of interest that the
configuration of this is opposite to that found in "primary" pros-

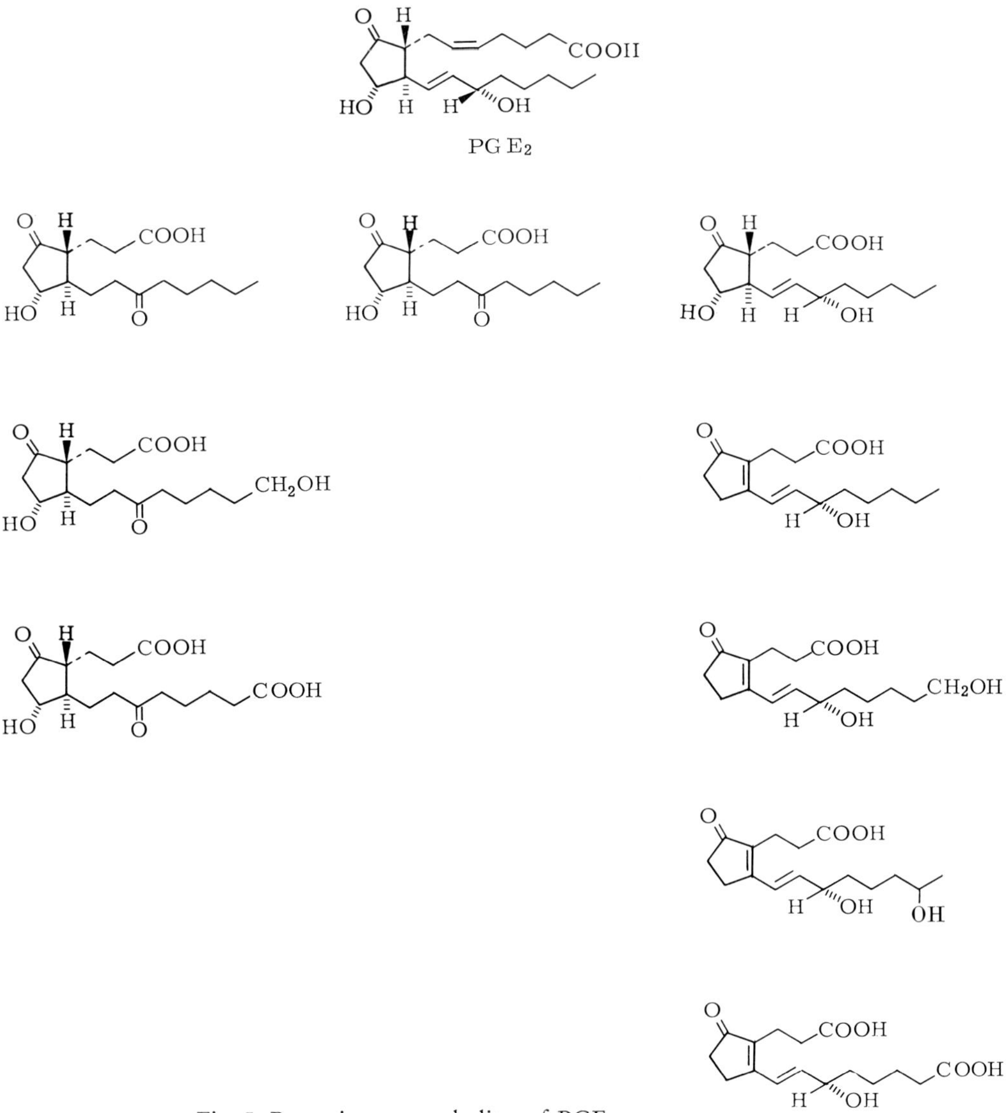

PG E$_2$

Fig. 5. Rat urinary metabolites of PGE$_2$

taglandins of the F series. The guinea-pig urinary metabolites of PGE$_2$ and PGF$_{2\alpha}$ thus differ only in the stereochemical configuration of the hydroxyl at the C-5 position (Fig. 7).

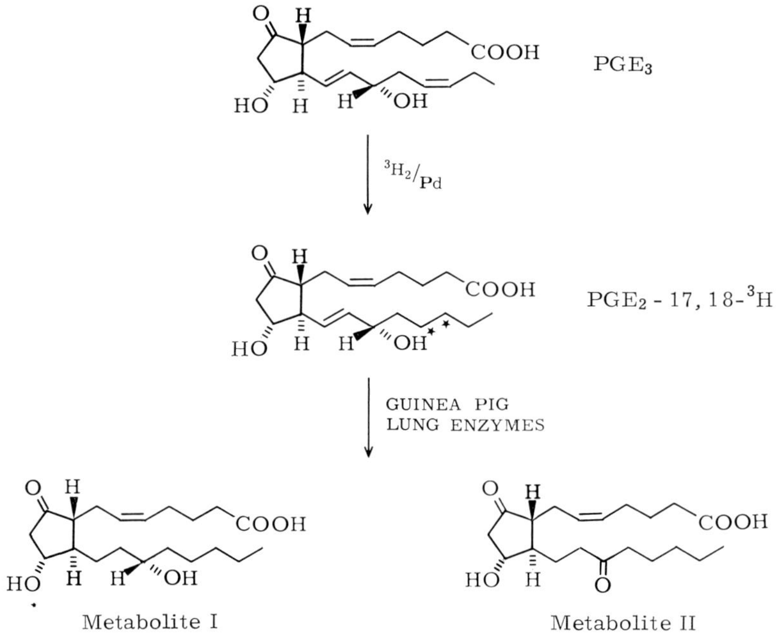

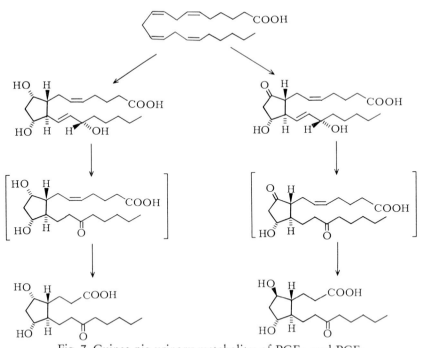

PGE₃

$^{3}H_{2}/Pd$

PGE$_2$ - 17, 18-^{3}H

GUINEA PIG
LUNG ENZYMES

Metabolite I

Metabolite II

Fig. 6. Guinea pig lung metabolites of PGE$_2$

Fig. 7. Guinea pig urinary metabolites of PGF$_{2\alpha}$ and PGE$_2$

5. Removal of Prostaglandins from the Circulation

Several workers have noted that the fall in arterial blood pressure with PGE_1 and PGE_2 is greater after intraaortic than after intravenous injection. This has been observed in sheep (HORTON, MAIN and THOMPSON, 1965), dogs (CARLSON and ORÖ, 1966), cats and rabbits (FERREIRA and VANE, 1967). However it was FERREIRA and VANE who established by using the blood-bathed organ technique (see Chapter II) that in the cat, dog, and rabbit, PGE_1, PGE_2 and $PGF_{2\alpha}$, although stable in blood, are 95% removed on one circulation through the lungs. Furthermore on infusion into the portal vein 80% of these prostaglandins are removed by the liver. VANE (1969) concludes that the body has a very efficient mechanism for preventing prostaglandins from reaching the arterial circulation, although prostaglandins released from the lungs (as in anaphylactic shock) would reach the arterioles and other target organs.

Independently, HORTON and JONES (1969) and McGIFF, TERRAGNO, STRAND, LEE, LONIGRO and NG (1969) demonstrated that PGA_1 and PGA_2, unlike PGE and PGF compounds, are not removed on passage through the lungs of the cat and dog, although they are metabolised by the Krebs-perfused guinea-pig lung (PIPER, VANE and WYLLIE, 1970). PGA_1 and PGA_2 are fairly effectively removed by the liver (55—90%) (HORTON and JONES, 1969).

Thus any PGA released from a tissue into the venous effluent could reach target organs before substantial inactivation had occured. The inactivation of PGE and PGF by the lungs may be thought to preclude a truly hormonal role. However this argument does not apply to PGA compounds which may be considered as potential circulating hormones.

6. Prostaglandin Dehydrogenase

Homogenates of pig lung contain an enzyme system which converts PGE_1 to its 15-oxo derivative (ÄNGGÅRD and SAMUELSSON, 1966 a). The enzyme, 15-hydroxy-prostaglandin dehydrogenase has been purified from high-speed supernatant fractions by ammonium sulphate fractionation and chromatography on TEAE-cellulose,

DEAE-Sephadex and Sephadex G-100. An 11-fold purification with a 30% yield was obtained.

The purified enzyme is NAD^+-dependent and is highly specific for the C-15 secondary alcohol substituent of the prostaglandins. The enzyme is denatured by temperatures above 55° and the reaction rate is slow at 30°. Most assays have been performed at 44° and within the pH range 6 to 8. PGE, PGF, PGA, α-nor-E and ω-homo-E compounds are all substrates for the enzyme, and a highly sensitive method for the determination of these compounds has been based upon this system by ÄNGGÅRD (see Chapter II). PGB compounds and their 19-hydroxy derivatives are not substrates; nor are any of the non-prostaglandin hydroxyl-containing compounds which have been tested. The enzyme is also stereospecific with regard to the configuration at C-15 (NAKANO, ÄNGGÅRD and SAMUELSSON, 1969; SHIO, ANDERSEN, COREY and RAMWELL, 1969). Both PGB compounds and 15-epi-PGE_1 are non-competitive inhibitors of the enzyme. Dihydro-PGE_1 is not a good substrate for the enzyme and so it is probable that *in vivo* oxidation of the secondary alcohol at C-15 precedes reduction of the C-13,14 double bond (SAMUELSSON et al., 1971).

Prostaglandin dehydrogenase is widely distributed in the pig. The highest activities have been found in lung, spleen and kidney. Lower activities have been observed in the stomach, testis, liver and small intestine. The renal cortex contains about three times more enzyme than the medulla (SAMUELSSON et al., 1971). The enzyme has also been identified in the lungs of several species.

NISSEN and ANDERSEN (1968) developed a histochemical method for the localisation of prostaglandin dehydrogenase in tissue sections. The tissue is incubated at 37° with PGE_2 and its co-enzyme, NAD. The $NADH_2$ cytochrome reductase activity was examined after 1 to 15 min. These workers showed pronounced enzyme activity in the thick ascending limb of the loop of Henle and in the distal tubule of the rat (Fig. 8). A short period of hydration in dehydrated animals resulted in a decrease in prostaglandin dehydrogenase activity (NISSEN and ANDERSEN, 1969).

Histochemical studies have also revealed the presence of this enzyme in the Purkinje cell layer of the cerebellar cortex (SIGGINS, HOFFER and BLOOM, 1971). This may reflect a physiological role of prostaglandin in this region.

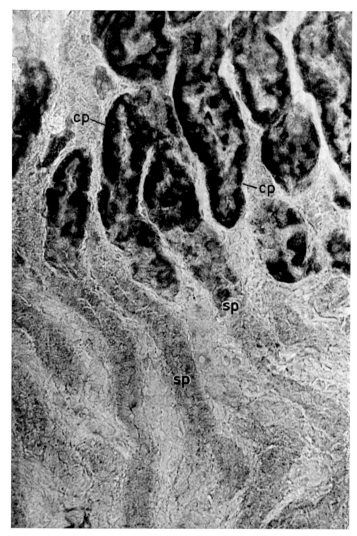

Fig. 8. Proximal tubules on the borderline between the straight portion (sp) and the distal part of the convoluted portion (cp) demonstrating prostaglandin dehydrogenase activity in the latter (magnification 1000). Rat kidney. (NISSEN and ANDERSEN, 1969)

Further work on the distribution of prostaglandin dehydrogenase in cells may contribute to our understanding of the physiological importance of the prostaglandins. Since 90—95% of PGE and PGF compounds are removed by one circulation through the lungs and since they are there rapidly oxidized and inactivated, the lung prostaglandin dehydrogenase clearly plays an important role in removing some biologically-active prostaglandins from the circulation. Moreover, in view of its high specificity, its widespread distribution and its relatively inactive end-products, (ÄNGGÅRD and SAMUELSSON, 1966 b), prostaglandin dehydrogenase may prove to be the most important enzyme for the inactivation of prostaglandins E and F produced locally in tissues.

Synthetic analogues of prostaglandins which possess biological activity but are not substrates for the enzyme could be of considerable practical importance. Furthermore compounds which block this enzyme system competitively would be valuable both as research tools and possibly also for clinical purposes.

7. Prostaglandin Reductase

This enzyme which reduces the C-13,14 double bond, was first reported in guinea-pig lung (ÄNGGÅRD and SAMUELSSON, 1964). The richest source is pig adipose tissue, in contrast to pig lung, which does not contain significant amounts of the enzyme. The reductase is also abundant in the spleen, kidney, liver, adrenals and small intestine (SAMUELSSON et al., 1971).

Like prostaglandin dehydrogenase, the reductase is located mainly in the particle-free fraction on subcellular fractionation.

The substrate specificity of this enzyme has not yet been reported. It is known however that the products of the reaction may retain considerable biological activity, for example, 13,14-dihydro E_1 has 14 to 35% of the activity of PGE_1 on isolated smooth muscle and is more active on the blood pressure (ÄNGGÅRD and SAMUELSSON, 1966 b).

8. Beta-oxidation

Many prostaglandins are substrates for the β-oxidizing enzyme system of rat liver mitochondria. Gas chromatographic-mass spectrometric analysis of the metabolites has shown that the carboxyl side chain is degraded by either one or two two-carbon fragments. However C-16 derivatives with two carbon atoms between the carboxyl group and the five membered ring are not subject to β-oxidation. It is of particular interest that whereas PGE_1 is oxidized only to the C-18 derivative, its fully saturated analogue $11\alpha,15\alpha$-dihydroxy-9-oxo-prostanoic acid is degraded one step further to the C-16 derivative. Both C-16 and C-18 derivatives were formed when 13,14-dihydro-PGE_1 and 11α-hydroxy-9,15-di-oxo-prostanoic acid (the lung metabolites) were incubated with the enzyme system (HAMBERG, 1968).

9. Omega-oxidation

Derivatives of PGA and PGB hydroxylated at C-19 are natural constituents of human seminal plasma (HAMBERG and SAMUELSSON, 1965). 19-hydroxy-PGA_1 and 20-hydroxy-PGA_1 are formed when PGA_1 is incubated with the microsomal fraction of guinea-pig and human liver (Fig. 9) (ISRAELSON, HAMBERG and SAMUELSSON, 1969;

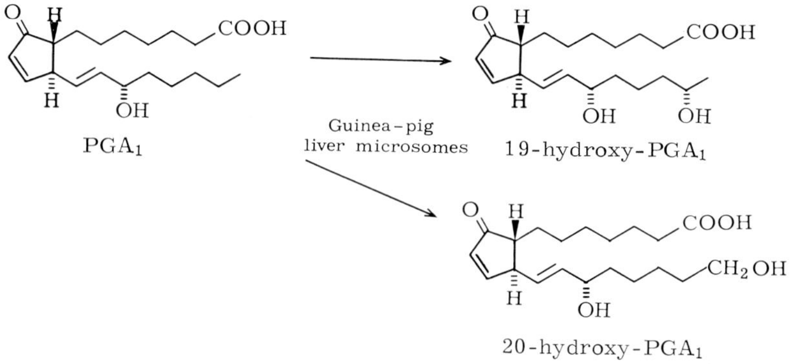

Fig. 9. Hydroxylated PGA_1 metabolites produced by guinea pig microsomal ω-oxidation

SAMUELSSON et al., 1971). PGE_1 is not a substrate for this microsomal ω-oxidising enzyme, nor could the enzyme be detected in rat liver microsomes. However ω-oxidation is probably less restricted than these findings might suggest since ω-carboxylated metabolites of PGE and PGF have been found in the urine of rat, guinea-pig and man.

10. Prostaglandin Isomerase

Incubation of PGE_1 with whole blood for up to 3 hours results in no loss of biological activity (HOLMES, HORTON and STEWART, 1968; JONES, 1970 a). In contrast PGA_1 and PGA_2 lose about 50% of their activity in 30 minutes on incubation with cat plasma (JONES, 1970 b) (Fig. 10).

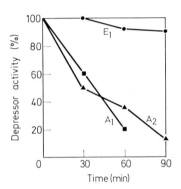

Fig. 10. Incubation of PGE_1, PGA_1 and PGA_2 with cat plasma, depressor activity of the incubation mixture estimated at intervals by intravenous injection into a cat. Ordinate: % loss in depressor activity. Abscissa: time in minutes. (JONES, 1970 b)

A high proportion of the PGA compound is converted (isomerised) to the corresponding PGB (Fig. 11) by an enzyme, prostaglandin isomerase, which is found in the plasma of several species (cat, dog, rabbit and pig but not in man or guinea-pig). The enzyme system has a pH optimum about 8.5 (Fig. 12) and is inactivated by heating at 70° but not at 55° (Fig. 13) (JONES, 1970 b).

The enzyme activity may be assayed conveniently by incubating with PGA_1 as substrate and measuring the U.V. absorption at 280 nm (Fig. 14). There is preliminary evidence that a co-factor (present in plasma) is required but this has so far not been identified.

This enzyme system may be of physiological significance in inactivating PGA compounds in the circulation, since these unlike PGE and PGF are not removed by the lungs (HORTON and JONES, 1969).

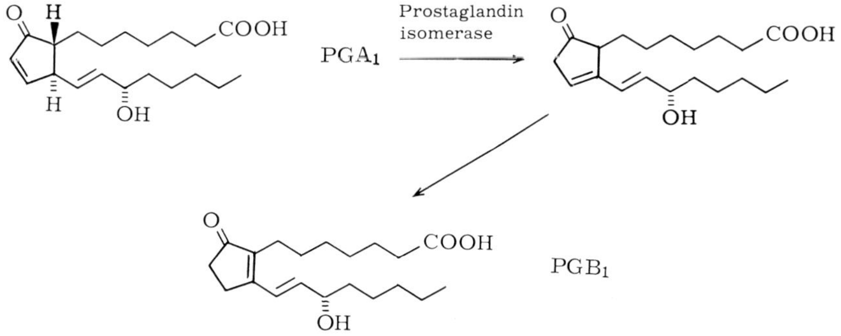

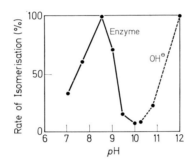

Fig. 11. Conversion of PGA₁ to PGB₁ by prostaglandin isomerase

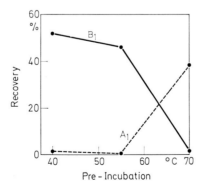

Fig. 12. Effect of pH on the conversion of PGA to PGB in the presence and absence (effect of base alone) of prostaglandin isomerase. (JONES, 1970 b)

Fig. 13. The effect of heat (40°, 55° and 70°) pre-treatment on the enzymatic activity of prostaglandin isomerase. (JONES, 1970 b)

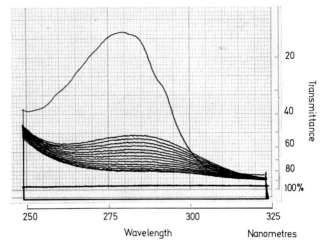

Fig. 14. The increase in U.V. absorption at 280 nm with time of incubation of PGA₁ in the presence of prostaglandin isomerase (JONES, 1970 b)

11. Human Metabolism

When labelled PGE₂ is injected intravenously in humans, about
50% of the radioactivity is recovered in the urine within 5 hours;
during the following 12 hours only a further 3% is excreted. One
urinary metabolite of PGE₂ in man has been identified as 7α-hy-
droxy-5,11-di-oxo-tetranor-prosta-1,16-dioic acid (Fig. 15) (HAMBERG
and SAMUELSSON, 1969 b).

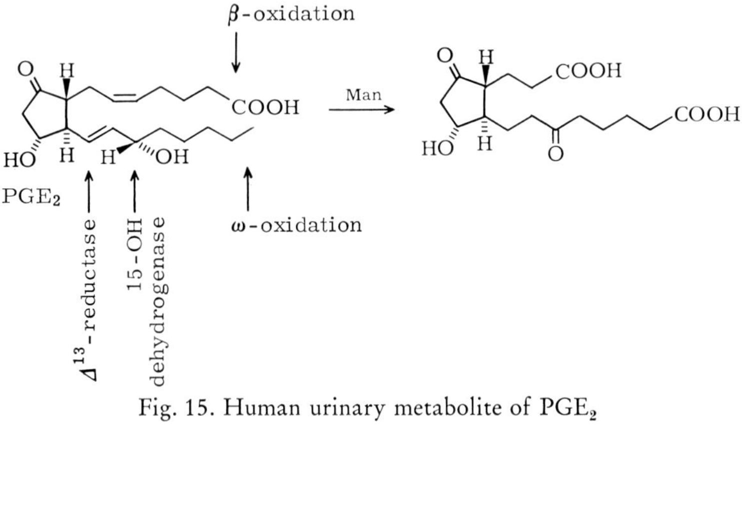

Fig. 15. Human urinary metabolite of PGE₂

The reaction sequence in the formation of this metabolite is likely to be as follows: oxidation of the C-15 by prostaglandin dehydrogenase (particularly in the lungs), reduction of the C-13,14 double bond by prostaglandin reductase, two steps of β-oxidation and ω-oxidation. The order of the last two steps is not established, but the first two reactions must precede β-oxidation since the tetranorprostaglandins are poor substrates for prostaglandin dehydrogenase.

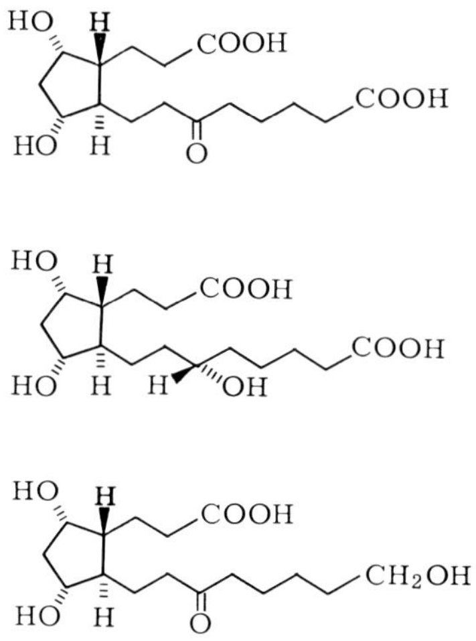

Fig. 16. Human urinary metabolites of PGF$_{2\alpha}$

Some indication of the rate of these reactions in man is given by the recent work of SAMUELSSON and his colleagues. One minute after the intravenous injection of tritiated PGE$_2$, a venous blood sample collected from the other arm was found to contain PGE$_2$ and 11α-hydroxy-9,15-di-oxo-5-cis-prostenoic acid in a ratio of about one to ten. The results suggested that about 4% of the original PGE$_2$ was present unchanged in the circulation after one minute and that 40% was present in the blood as the metabolite. From the work of Vane it seems very probable that a high proportion of the PGE$_2$ would be taken up by the lungs and from the work of ÄNGGÅRD and SAMUELSSON, we know that C-15 oxidation followed by C-13,14 reduction would occur. Inactivation of administered and circulating PGE$_2$ is thus a very rapid process. Subsequent steps in the metabolism which

6*

are less prostaglandin-specific probably occur in the liver and other tissues.

When labelled $PGF_{2\alpha}$ is injected intravenously in man, over 90% of the radioactivity is recovered from the urine within about 5 hours. The main metabolite, $5\alpha,7\alpha$-dihydroxy-11-oxo-tetranorprosta-1,16-dioic acid corresponds to the urinary metabolite of PGE_2 (Fig. 16) (GRANSTRÖM and SAMUELSSON, 1969 b).

Additional urinary metabolites of $PGF_{2\alpha}$ in man have been identified as $5\alpha,7\alpha,11$-trihydroxy-tetranorprosta-1,16-dioic acid and $5\alpha,7\alpha,16$-trihydroxy-11-oxo-tetranorprostanoic acid (Fig. 16). The delta lactones of two of the metabolites have also been found.

PGE_1 and $PGF_{1\alpha}$ give rise to the same main metabolites as PGE_2 and $PGF_{2\alpha}$ respectively. Thus the estimation of these compounds in urine will reflect the total production of the four prostaglandins of the E and F series.

No information is yet available about the metabolism of PGA and PGB compounds in man (or their urinary metabolites in other animals). The possibility that some PGE may be converted to PGA and PGB must be considered. For example tetranor-PGE_1 in the rat is metabolised by dehydration and not by oxidation.

SAMUELSSON et al. (1971) have estimated from measurements of urinary metabolites that the production (or "secretion") of PGE ranges from 109 to 226 µg/24 hours in males (4 subjects) and 23 to 48 µg/24 hours in females (2 subjects). These results need to be extended but they do indicate that relatively small quantities of prostaglandins are being synthesised. Evidently, a very small proportion of the precursor essential fatty acids consumed in the diet (of the order of 10 g linoleic acid/24 hours — Documenta Geigy, 1970) is converted to PGE or PGF.

References

ÄNGGÅRD, E., GRÉEN, K., SAMUELSSON, B.: Synthesis of tritium labelled prostaglandin E_2 and studies on its metabolism in guinea-pig lung. J. biol. Chem. 240, 1932—1940 (1965).
— SAMUELSSON, B.: Metabolism of prostaglandin E_1 in guinea pig lung: the structure of two metabolites. J. biol. Chem. 239, 4097—4102 (1964).
— — The metabolism of prostaglandin E_3 in guinea pig lung. Biochemistry, N. Y. 4, 1864—1871 (1965).

ÄNGGÅRD, E., SAMUELSSON, B.: Purification and properties of a 15-hydroxy-prostaglandin dehydrogenase from swine lung. Ark. Kemi **25**, 293—300 (1966 a).

— — Metabolites of prostaglandins and their biological properties. Mem. Soc. Endocr. **14**, 107—117 (1966 b).

CARLSON, L. A., ORÖ, L.: Effects of prostaglandin E_1 on blood pressure and heart rate in the dog. Acta physiol. scand. **67**, 89—99 (1966).

DAWSON, W., JESSUP, S. J., MCDONALD-GIBSON, W., RAMWELL, P. W., SHAW, J. E.: Prostaglandin uptake and metabolism by the perfused rat liver. Br. J. Pharmac. **39**, 585—598 (1970).

FERREIRA, S. H., VANE, J. R.: Prostaglandins: their disappearance from and release into the circulation. Nature **216**, 868—873 (1967).

GRANSTRÖM, E., INGER, U., SAMUELSSON, B.: The structure of a urinary metabolite of prostaglandin $F_{1\alpha}$ in the rat. J. biol. Chem. **240**, 457—461 (1965).

— SAMUELSSON, B.: The structure of the main urinary metabolite of prostaglandin $F_{2\alpha}$ in the guinea pig. Eur. J. Biochem. **10**, 411—418 (1969 a).

— — The structure of a urinary metabolite of prostaglandin $F_{2\alpha}$ in man. J. Amer. chem. Soc. **91**, 3398—3400 (1969 b).

GRÉEN, K.: Structures of urinary metabolites of prostaglandin $F_{2\alpha}$ in the rat. Acta chem. scand. **23**, 1453—1455 (1969).

— HANSSON, E., SAMUELSSON, B.: Synthesis of tritium-labelled prostaglandin $F_{2\alpha}$ and studies of its distribution by autoradiography. Prog. biochem. Pharmac. **3**, 85—88 (1967).

— SAMUELSSON, B.: On the excretion of dinor-prostaglandin $F_{1\alpha}$ in the rat. Prostaglandin Symposium of the Worcester Foundation for Exp. Biol. Eds.: P. W. RAMWELL and J. E. SHAW. New York: Interscience 1968, pp. 389—394.

HAMBERG, M.: Metabolism of prostaglandins in rat liver mitochondria. Eur. J. Biochem. **6**, 135—146 (1968).

— SAMUELSSON, B.: Isolation and structure of a new prostaglandin from human seminal plasma. Biochim. biophys. Acta **106**, 215—217 (1965).

— — The structure of a urinary metabolite of prostaglandin E_2 in the guinea-pig. Biochem. biophys. Res. Comm. **34**, 22—27 (1969 a).

— — The structure of the major urinary metabolite of prostaglandin E_2 in man. J. Amer. chem. Soc. **91**, 2177—2178 (1969 b).

HANSSON, E., SAMUELSSON, B.: Autoradiographic distribution studies of ^{3}H-labeled prostaglandin E_1 in mice. Biochim. biophys. Acta **106**, 379—385 (1965).

HOLMES, S. W., HORTON, E. W., STEWART, M. J.: Observations on the extraction of prostaglandins from blood. Life Sci. **7**, 349—354 (1968).

HORTON, E. W., JONES, R. L.: Prostaglandins A_1, A_2 and 19-hydroxy-A_1; their actions on smooth muscle and their inactivation on passage through the pulmonary and hepatic portal vascular beds. Brit. J. Pharmac. **37**, 705—722 (1969).

HORTON, E. W., MAIN, I. H. M., THOMPSON, C. J.: Effects of prostaglandins on the oviduct studied in rabbits and ewes. J. Physiol. (Lond.) **180**, 514—528 (1965).

ISRAELSSON, U., HAMBERG, M., SAMUELSSON, B.: Biosynthesis of 19-hydroxy-prostaglandin A_1. Eur. J. Biochem. **11**, 390—394 (1969).

JONES, R. L.: A prostaglandin isomerase in cat plasma. Biochem. J. **119**, 64—65 P (1970 a).

— Pharmacology of prostaglandins A and B. Ph.D. Thesis. University of London 1970 b.

McGIFF, J. C., TERRAGNO, N. A., STRAND, J. C., LEE, J. B., LONIGRO, A. J., NG, K. K. F.: Selective passage of prostaglandins across the lung. Nature (Lond.) **223**, 742—745 (1969).

NAKANO, J. R., ÄNGGÅRD, E., SAMUELSSON, B.: 15-hydroxy-prostanoate dehydrogenase. Prostaglandins as substrates and inhibitors. Eur. J. Biochem. **11**, 386—389 (1969).

NISSEN, H. M., ANDERSEN, H.: On the localization of a prostaglandin-dehydrogenase activity in the kidney. Histochemie **14**, 189—200 (1968).

— — On the activity of a prostaglandin dehydrogenase system in the kidney. A histochemical study during hydration/dehydration and salt-repletion/salt depletion. Histochemie **17**, 241—247 (1969).

PIPER, P. J., VANE, J. R., WYLLIE, J. H.: Inactivation of prostaglandins by the lungs. Nature (Lond.) **225**, 600—604 (1970).

SAMUELSSON, B.: Synthesis of tritium-labeled prostaglandin E_1 and studies on its distribution and excretion in the rat. J. biol. Chem. **239**, 4091—4096 (1964).

— GRANSTRÖM, E., GRÉEN, K., HAMBERG, M.: Metabolism of prostaglandins. Ann. N. Y. Acad. Sci. **180**, 138—161 (1971).

SHIO, H., RAMWELL, P. W., ANDERSEN, N. H., COREY, E. J.: Stereo-specificity of the prostaglandin 15-dehydrogenase from swine lung. Experientia **26**, 355—357 (1970).

SIGGINS, G. R., HOFFER, B. J., BLOOM, F. E.: Specificity of prostaglandin E_1 — norepinephrine antagonisms in the brain: microelectrophoretic and histochemical correlations. Ann. N. Y. Acad. Sci. **180**, 302—319 (1971).

VANE, J. R.: The release and fate of vasoactive hormones in the circulation. Br. J. Pharmac. **35**, 209—242 (1969).

VONKEMAN, H., NUGTEREN, D. H., VAN DORP, D. A.: The action of prostaglandin 15-hydroxydehydrogenase on various prostaglandins. Biochim. biophys. Acta **187**, 581—583 (1969).

VI. Female Reproductive Tract Smooth Muscle

1. Parturition

The stimulant action of human semen (0.25—1 ml) on the non-pregnant uterus when introduced into the vagina or uterine lumen was reported by KARLSON (1959). Increased motility of the body of the uterus was accompanied by decreased motility of the cervix, pressure measurements being made by a method in which transducers are inserted into the fundus, isthmus and cervix. In the light of subsequent investigations it is almost certain that these effects can be attributed to the presence of prostaglandins in semen.

ELIASSON and POSSE (1960) extracted prostaglandins from human seminal plasma and administered these intravaginally to volunteers. They observed an increase in uterine motility in five out of seven subjects at the estimated time of ovulation. At other phases in the

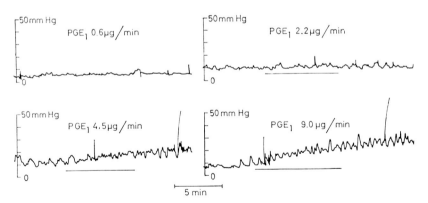

Fig. 1. The effect of stepwise increased doses of PGE_1 on the motility of the midpregnant human uterus (eighteenth week), recorded by measuring the amniotic pressure. PGE_1 was administered by continuous intravenous infusion. The infusion was started at the beginning of each curve section illustrated. The horizontal lines represent original uterine tone (BYGDEMAN et al., 1967)

menstrual cycle (menstruation, early proliferative, late secretory) no effects on uterine motility were detected.

The pregnant human myometrium is especially sensitive to prostaglandins. Thus BYGDEMAN, KWON and WIQVIST (1967) recorded an increase in amplitude and frequency of contractions in women at mid-pregnancy with intravenous infusions of PGE_1 (0.6—9.0 µg/min) (Figs. 1 and 2). BYGDEMAN and his co-workers (BYGDEMAN,

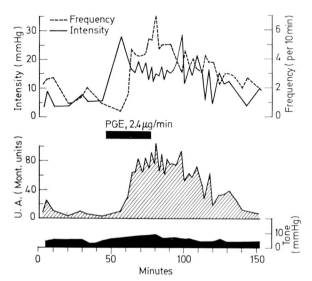

Fig. 2. The effect of intravenous infusion of 2.4 µg of PGE_1 per minute in the thirty-second week of pregnancy. The tracing is analyzed in accordance with the Montevideo method. U.A. = Uterine activity (BYGDEMAN, KWON, MUKHERJEE and WIQVIST, 1968)

KWON, MUKHERJEE and WIQVIST, 1968; WIQVIST, BYGDEMAN, KWON, MUKHERJEE and ROTH-BRANDEL, 1968; BYGDEMAN, KWON, MUKHERJEE, ROTH-BRANDEL and WIQVIST, 1970) have extended these observations to $PGF_{2\alpha}$ (Fig. 3 and 4) and PGE_2, and have shown that intravenous infusions of these prostaglandins induce abortion (ROTH-BRANDEL, BYGDEMAN, WIQVIST and BERGSTRÖM, 1970).

Following his discovery of prostaglandins in umbilical cord vessels, in amniotic fluid and finally in the circulation of women in labour (KARIM, 1966, 1967; KARIM and DEVLIN, 1967; KARIM, 1968), KARIM (1969) put forward the suggestion that $PGF_{2\alpha}$ might be impli-

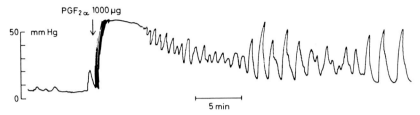

Fig. 3. Maximum elevation of uterine tone followed by labour-like activity after intravenous injection of 1,000 μg PGF$_{2α}$ (eighteenth week of pregnancy) (BYGDEMAN, KWON, MUKHERJEE, ROTH-BRANDEL and WIQVIST, 1970)

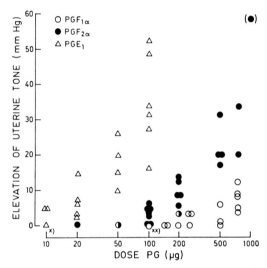

Fig. 4. Maximum rise of uterine tone (above resting tone level) following single intravenous injections of PGE$_1$, PGF$_{2α}$, and PGF$_{1α}$. The data were obtained from 13 midpregnant women: (×) represents 4 injections given to 3 patients; (××) represents 5 injections given to 5 patients. (●) 1,000 μg PGF$_{2α}$ caused a rise of tone beyond the measurable range (BYGDEMAN et al., 1970)

cated physiologically in the uterine contractions of parturition. Moreover he showed that this compound can be used to induce labour (Table 1). His first report of the successful use of PGF$_{2α}$ (Fig. 5) for this purpose (KARIM, TRUSSELL, PATEL and HILLIER, 1968) has been confirmed by several groups (WIQVIST and BYGDEMAN, 1970; KARIM, TRUSSELL, HILLIER and PATEL, 1969) and several reports on this

Table 1. *Results of infusion of prostaglandin F$_{2\alpha}$ in pregnant women at or near term*

No.	Age	Gravida	Maturity	Indication	Infusion time (hr)	Cervical Dilatation		Infusion — delivery time (hr)	Weight (lb)	Result
						At start of infusion (cm)	At finish of infusion (cm)			
1	22	4	36	I.U.D.	6.0	0	6	10.0	5.6	Stillbirth
2	28	6	39	Prolapse of cord +I.U.D.	4.0	2	8	6.0	8.0	Stillbirth
3	28	7	44	Postmaturity	2.2	2	6	6.5	8.0	Alive
4	34	11	43	Postmaturity	3.5	2	8	4.9	7.0	Alive
5	28	9	34	Premature rupture of membranes	2.0	3	8	5.5	4.6	Alive
6	16	1	44	Postmaturity	3.0	0	5	12.5	8.6	Alive
7	31	6	40	Uterine inertia	2.5	3	8	2.7	5.8	Alive
8	26	7	43	Postmaturity	1.0	2	8	2.0	6.5	Alive
9	21	2	44	Postmaturity	{ 5.0 ; 1.5	0	3 ; 6	7.0 [a] ; 7.0	7.0	Alive
10	22	2	38	Premature rupture of membranes	3.5	2	6	8.3	8.4	Alive

The table summarises the results of ten cases. In all cases the rate of infusion did not exceed 0.05 µg/kg/min. In cases 3 and 10 the rate infusion was reduced to 0.025 µg/kg/min after 1¼ hours and 2¼ hours respectively. One patient (case No. 7) was an Asian; the remaining nine were African Negro. I.U.D.=intra-uterine death.

[a] Stopped contracting after 7 hr.

topic have now been published (EMBREY, 1970 a; BEAZLEY, DEWHURST
and GILLESPIE, 1970). Moreover, PGE_2 (Fig. 6) proves to be even
more active than $PGF_{2\alpha}$ (KARIM, HILLIER, TRUSSELL, PATEL and
TAMUSANGE, 1970). The final assessment of the clinical value and
safety of prostaglandins for the induction of parturition must await
the outcome of controlled clinical trials now in progress.

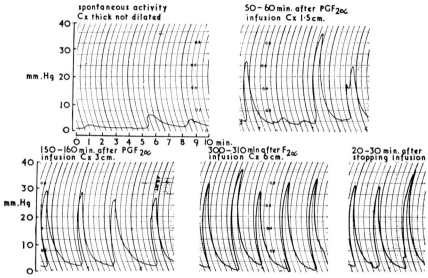

Fig. 5. Effect of prostaglandin $F_{2\alpha}$ infusion at 0.05 μg/kg/min on uterine
activity of pregnant woman (Case 1) with dead foetus at 36 weeks gestation. The infusion was stopped after six hours when the cervix had dilated
6 cm. Membranes ruptured four and a half hours after the start of the prostaglandin infusion, and the dead foetus was delivered 10 hours after the
infusion was begun. Cx = Cervical dilatation (KARIM, TRUSSELL, PATEL
and HILLIER, 1968)

2. Abortion

KARIM and FILSHIE (1970 a) published the first detailed report of
the use of intravenous $PGF_{2\alpha}$ (50 μg/min) for the induction of abortion between the 9th and 22nd week of pregnancy.

Stimulation of uterine activity was observed. In 10 out of 15
women abortion was complete, the products of conception (foetus
and placenta) being simultaneously expelled between 4 and 27 hours

(induction-abortion interval). In three cases placental expulsion occurred after a further 3 to 7 hours. In only one case was surgical removal of the placenta required and in one case the women did not abort in spite of good tone and frequent uterine contractions (Table 2).

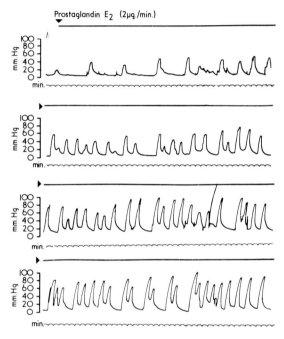

Fig. 6. Intravenous infusion of PGE$_2$ for induction of labour. Sample records taken 0, 1, 2.5 and 3.5 hours after start of infusion (EMBREY, 1970 a)

BYGDEMAN and WIQVIST (1971) administered PGF$_{2\alpha}$ intravenously to 69 women during the 6th to 20th week of gestation. The infusion rate (25—100 µg/min) was increased stepwise until slight subjective side effects appeared. The dose was then adjusted to the highest rate acceptable to the individual subject. With an infusion period of 7 hours and a total dose of 30 mg, 94% of the early pregnancies (up to 8 weeks duration) were terminated. Only 10 to 30% of later stages of gestation were terminated even when the infusion period was doubled. These results show a lower success rate than KARIM and FILSHIE (1970 a) possibly because the length of infusion was shorter.

Table 2. *Details cases of therapeutic abortion induced by infusion of 50 µg per min prostaglandin $F_{2\alpha}$* (KARIM and FILSHIE, 1970 a)

No.	Age	Nationality	Para	Gravida	Gestation (wk.)	Induction-abortion interval	Interval between the delivery of the foetus and placenta	Remarks
1	21	British	1	1	16	12 hr 15 min	0 min	
2	16	West Indian	0	1	17	24 hr 30 min	0 min	Previous hysterotomy
3	29	British	6	7	14	6 hr	3 hr	
4	20	Nigerian	3	3	14	10 hr	0 min	Diarrhoea
5	28	Nigerian	5	6	19	6 hr 20 min	0 min	Diarrhoea and vomiting
6	22	British	0	0	14	17 hr	0 min	Diarrhoea and vomiting
7	17	West Indian	1	1	16	4 hr 20 min	0 min	Diarrhoea
8	27	British	4	4	12	8 hr 15 min	—	Evacuation of retained products of conception; diarrhoea
9	35	West Indian	4	7	15	9 hr	7 hr	Diarrhoea
10	20	British	3	3	9	12 hr 40 min	7 hr	—
11	24	West Indian	2	3	22	17 hr 47 min	0 min	—
12	36	British	5	5	12	11 hr 30 min	0 min	—
13	31	British	0	0	16	Failed*	—	
14	18	British	0	0	12	8 hr	0 min	
15	22	West Indian	4	4	16	27 hr 15 min	0 min	Diarrhoea and vomiting

* Failed induction after 48 hours infusion.

Untoward side effects reported by BYGDEMAN and WIQVIST (1971) during $PGF_{2\alpha}$ infusion include dysmenorrhoea (50% at 50 µg/min and 100% incidence at 100 µg/min), nausea and diarrhoea (12% at 50 µg/min and 30% incidence at 75 µg/min).

PGE_2 (Fig. 7) and PGE_1 have also been used with success in termination of pregnancy (KARIM and FILSHIE, 1970 b; EMBREY, 1970 b; BYGDEMAN and WIQVIST, 1971). The total dose of PGE_2 required varies between 1 and 10 mg infused over 6 to 12 hours. The dose of $PGF_{2\alpha}$ is higher (10—100 mg) infused over a similar period.

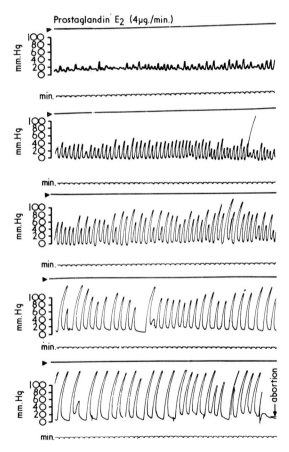

Fig. 7. Intravenous infusion of PGE_2 for induction of abortion. Sample records begin at 0.5, 4, 9, 12.5 and 17.5 hours after start of the infusion (EMBREY, 1970 b)

The optimal formulation, dosage and method of administration have still to be established.

Other routes of administration are also successful. In order to use PGE_2 or $PGF_{2\alpha}$ for induction of abortion on an out-patient basis, BYGDEMAN and WIQVIST (1971) have administered these compounds into the uterine cavity between the foetal membranes and the uterine wall. The prostaglandin was administered intermittently via an indwelling catheter. PGE_2 (25—75 µg) or $PGF_{2\alpha}$ (200—500 µg) at 1 to 2 hour intervals produced sustained and intensive contractions (Fig. 8). The total dose required was about one-tenth of that needed by intravenous infusion (Table 3). Neither nausea nor diarrhoea was experienced.

The possibility of using prostaglandins as contraceptives has also been investigated by BYGDEMAN and WIQVIST (1971). $PGF_2\alpha$ has been administered to women a few days following the first missed menstrual period. Bleeding started 2 to 3 hours following the beginning of infusion. The early results with this new method appear

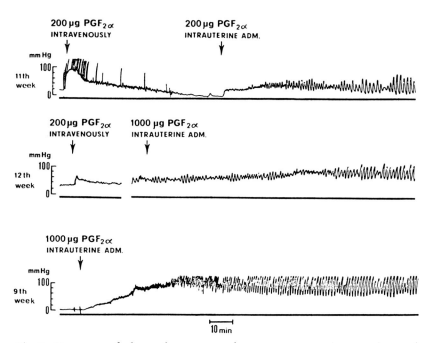

Fig. 8. Response of the early pregnant human uterus to intrauterine and intravenous administration of $PGF_{2\alpha}$ (BYGDEMAN and WIQVIST, 1971)

Table 3. *Intrauterine administration of prostaglandins for induction of therapeutic abortion* (BYGDEMAN and WIQVIST, 1971)

Week of pregnancy	Compound	Total dose (μg)	Interval between first and last injection (hours)	Complete or partial expulsion of the conceptus
5	$PGF_2\alpha$	2500	6.5	+
7	$PGF_2\alpha$	2700	6.3	+
7	$PGF_2\alpha$	950	7.3	+
7	$PGF_2\alpha$	1950	7.5	+
7	$PGF_2\alpha$	1500	4.3	+
7	$PGF_2\alpha$	1750	5.0	+
8	$PGF_2\alpha$	3200	9.3	+
8	$PGF_2\alpha$	5400	3.3	+
10	$PGF_2\alpha$	1000	2.5	+
7	PGE_2	350	6.3	+
8	PGE_2	25	[a]	+
13	PGE_2	1050	36.0	[b]

[a] Only one injection.
[b] Conceptus retained but cervix dilated.

to be promising and a number of women in Sweden unable or unwilling to use conventional methods now rely exclusively upon prostaglandins for contraceptive purposes.

There is also evidence that the intravaginal route is effective for the termination of pregnancy (KARIM, 1971). This might carry less danger of infection than the intrauterine method and seems to be similarly free of side effects. PGE_2 in 20 mg doses given via the vagina induces uterine contractions (Fig. 9). It is not yet clear whether this intravaginal PGE_2 is absorbed into the circulation or whether it exerts its effects primarily by a local action. Tritiated PGE_1 is absorbed from the human vagina (SANDBERG, INGELMAN-SUNDBERG, RYDÉN and JOELSSON, 1968) and the amounts may be sufficient to affect tubal motility in non-pregnant women as measured by Rubin's test (ELIASSON and POSSE, 1965). Evidence has been obtained from experiments in anaesthetised rabbits that PGE_1 can be absorbed from the lumen of the vagina in amounts sufficient to cause inhibition of tubal contractions (HORTON, MAIN and THOMPSON, 1965). Earlier ASPLUND (1947) had stated that "prostaglandin" is

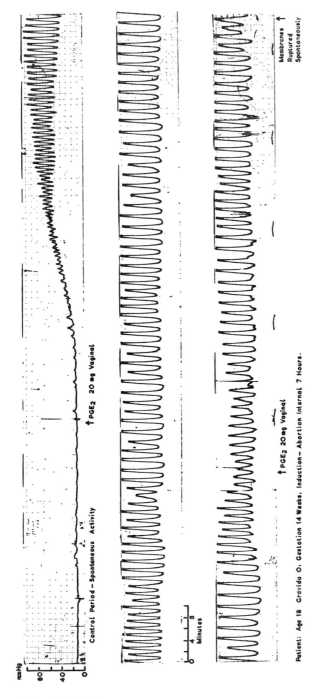

Fig. 9. Continuous record of the effect of the intra-vaginal administration of PGE_2 on the pregnant human uterus at 14 weeks (KARIM, 1971)

active upon rabbit tubal smooth muscle when administered by various routes including intravaginally.

No serious side-effects have been reported from the use of large doses of $PGF_{2\alpha}$ and PGE_2 infused intravenously to induce abortion. The higher doses of both prostaglandins (but particularly $PGF_2\alpha$) tend to produce diarrhoea and vomiting of mild severity. Following PGE_2 some sedation has been observed (FILSHIE, personal communication). There were no changes in pulse rate or arterial blood pressure. Controlled trials on the use of prostaglandins in the termination of pregnancy are in progress.

3. Human Myometrial Strips in vitro

Many observations have been made on the response of human myometrial strips *in vitro* to prostaglandins (BYGDEMAN, 1964, 1967; SANDBERG, INGELMAN-SUNDBERG and RYDÉN, 1963, 1964, 1965; ELIASSON, 1966). In general both amplitude and frequency of spontaneous contractions are *reduced* by both whole semen and by PGE compounds. The threshold concentration for PGE_1 and PGE_2 is about 10 to 100 ng/ml. PGA compounds and their 19-hydroxy derivatives, all natural constituents of human semen are also inhibitory but 10 to 30 times less potent than PGE_1 (BYGDEMAN, HAMBERG and SAMUELSSON, 1966; BYGDEMAN, 1967; BYGDEMAN and HAMBERG, 1967). $PGF_{1\alpha}$ and $PGF_{2\alpha}$ stimulate contractions of the isolated human myometrium. Preparations obtained late in the menstrual cycle or during pregnancy are very sensitive to this stimulant action of $PGF_{2\alpha}$ suggesting that hormonal status affects the sensitivity of these isolated strips. This is confirmed by the observation that the uterus is some 5 times more sensitive to the inhibitory action of PGE_1 at the time of ovulation. Moreover, the myometrial strip from a pregnant women is often contracted by PGE_1, though higher doses can cause inhibition (EMBREY, 1969; EMBREY and MORRISON, 1968).

4. Human Fallopian Tubes

PGE_1 and PGE_2 contract the most proximal (uterine) end of the human Fallopian tube *in vitro* but relax the distal three-quarters (SANDBERG et al., 1963, 1964). The difference is most marked in tissue removed during the secretory phase of the menstrual cycle. All parts of the tube are relaxed by PGE_3 and contracted by $PGF_{1\alpha}$ and $PGF_{2\alpha}$ (SANDBERG et al., 1964, 1965). All these prostaglandins occur in human semen and their action may result in the retention of the ovum within the tube and so increase the changes of fertilisation following coitus.

5. Uterus and Oviduct of Other Species

The response of female reproductive tract smooth muscle of several laboratory animals to PGE_1 has been investigated both *in vivo* and *in vitro*. Rat, guinea-pig and rabbit uterus *in vitro* contracts, guinea-pig uterus *in vivo* contracts, but both rabbit uterus and oviduct *in vivo* are inhibited (Fig. 10) by PGE_1 (BERGSTRÖM, ELIASSON, VON EULER and SJÖVALL, 1959; BERTI and NAIMZADA, 1965; HORTON, 1963; HAWKINS, JESSUP and RAMWELL, 1968; HORTON and MAIN, 1963, 1965, 1966; HORTON, MAIN and THOMPSON, 1965; SULLIVAN, 1966; CLEGG, HALL and PICKLES, 1965). The isolated rat

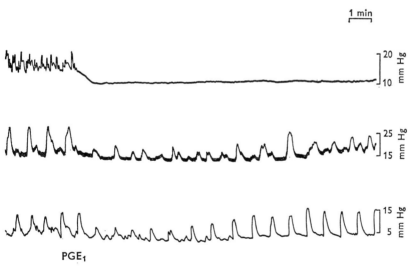

PGE₁

Fig. 10. Rabbit, 2·5 kg, anaesthetised with urethane (1.75 g/kg) injected intra-peritoneally. Records of intra-luminal pressure of the oviduct (upper trace), uterine horn (middle trace) and cervix (lower trace). PGE_1 = prostaglandin E_1 (6 µg) injected intravenously (HORTON, MAIN and THOMPSON, 1965)

uterus has been used as a bioassay preparation (see Chapter II). This tissue is more sensitive to $PGF_{1\alpha}$ and $PGF_{2\alpha}$ than PGE_1 and PGE_2 (HORTON and MAIN, 1963, 1965) but the sensitivity can be greatly increased by prior ovariectomy of the animal (HAWKINS et al., 1968). Marked tachyphylaxis has been observed on this tissue with several prostaglandins (ADAMSON, ELIASSON and WIKLUND, 1967; ELIASSON, BRZDEKIEWICZ and WIKLUND, 1969). Guinea-pig uterus *in vitro* is not only very sensitive to PGE compounds but also shows the phenomenon of enhancement to the action of other oxytocic substances and electrical stimulation (Fig. 11) (CLEGG et al., 1966). This could find an important clinical application if PGE_2 is found to potentiate the actions of oxytocin on the human myometrium.

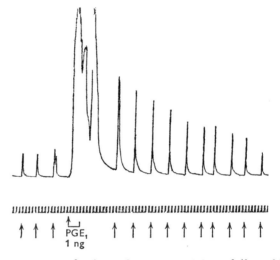

Fig. 11. Direct response of guinea-pig uterus to PGE_1, followed by enhancement of responses to stimulation (at arrows) by electric field 1 V/cm. Suspension medium contained K^+ 5.9 mM, Ca^{2+} 3 mM (CLEGG et al., 1966)

References

ADAMSON, U., ELIASSON, R., WIKLUND, B.: Tachyphylaxis in rat uterus to some prostaglandins. Acta physiol. scand. 70, 451—452 (1967).

ASPLUND, J.: Some preliminary experiments in connection with the effect of prostaglandin on the uterus and tubae *in vivo*. Acta physiol. scand. 13, 109—114 (1947).

BEAZLEY, J. M., DEWHURST, C. J., GILLESPIE, A.: Induction of labour with prostaglandin E_2. J. Obstet. Gynaec. Brit. Commonw. **77**, 193—199 (1970).

BERGSTRÖM, S., ELIASSON, R., VON EULER, U. S., SJÖVALL, J.: Some biological effects of two crystalline prostaglandin factors. Acta physiol. scand. **45**, 133—144 (1959).

BERTI, F., NAIMZADA, M.: Spiccata attività della prostaglandina E_1 (PGE$_1$) sull' utero di cavia. Boll Soc. ital. Biol. sper. **41**, 1324—1326 (1965).

BYGDEMAN, M.: The effect of different prostaglandins on the human myometrium *in vitro*. Acta physiol. scand. **63**, suppl. 242, 1—78 (1964).

— Studies of the effects of prostaglandins in seminal plasma on human myometrium in vitro. Nobel Symposium 2, Prostaglandins. Eds.: S. BERGSTRÖM and B. SAMUELSSON. Stockholm: Almqvist and Wiksell 1967, pp. 71—77.

— HAMBERG, M.: The effect of eight new prostaglandins on human myometrium. Acta physiol. scand. **69**, 320—326 (1967).

— — SAMUELSSON, B.: The content of different prostaglandins in human seminal fluid and their threshold doses on the human myometrium. Mem. Soc. Endocr. **14**, 49—64 (1966).

— KWON, S. U., MUKHERJEE, T., ROTH-BRANDEL, U., WIQVIST, N.: The effect of the prostaglandin F compounds on the contractility of the pregnant human uterus. Amer. J. Obstet Gynec. **106**, 567—572 (1970).

— — — WIQVIST, N.: Effect of intravenous infusion of prostaglandin E_1 and E_2 on motility of the pregnant human uterus. Amer. J. Obstet. Gynec. **102**, 317—326 (1968).

— — WIQVIST, N.: The effect of prostaglandin E_1 on human pregnant myometrium *in vivo*. Nobel Symposium 2, Prostaglandins. Eds.: S. BERGSTRÖM and B. SAMUELSSON. Stockholm: Almqvist and Wiksell 1967, pp. 93—96.

— WIQVIST, N.: Early abortion in the human. Ann. N. Y. Acad. Sci. **180**, 473—482 (1971).

CLEGG, P. C., HALL, W. J., PICKLES, V. R.: The action of ketonic prostaglandins on the guinea pig myometrium. J. Physiol. (Lond.) **183**, 123—144 (1966).

ELIASSON, R.: The effect of different prostaglandins on the motility of the human myometrium. Mem. Soc. Endocr. **14**, 77—88 (1966).

— BRZDEKIEWICZ, Z., WIKLUND, B.: Tachyphylactic response of the isolated rat myometrium to prostaglandins E. In: Prostaglandins, Peptides and Amines. Eds.: P. MANTEGAZZA and E. W. HORTON. London: Academic Press 1969, pp. 57—64.

— POSSE, N.: The effect of prostaglandin on the non-pregnant human uterus *in vivo*. Acta obstet. gynec. scand. **39**, 112—126 (1960).

— — Rubin's test before and after intravaginal application of prostaglandin. Int. J. Fert. **10**, 373—377 (1965).

EMBREY, M. P.: The effect of prostaglandins on the human pregnant uterus. J. Obstet. Gynaec. Brit. Commonw. **76**, 783—789 (1969).

EMBREY, M. P.: Induction of labour by prostaglandins E_1 and E_2. Brit. med. J. 2, 256—258 (1970 a).
— Induction of abortion by prostaglandins E_1 and E_2. Brit. med. J. 2, 258—260 (1970 b).
— MORRISON, D. L.: The effect of prostaglandins on human pregnant myometrium *in vitro*. J. Obstet. Gynaec. Brit. Commonw. 75, 829—832 (1968).
HAWKINS, R. A., JESSUP, R., RAMWELL, P. W.: Effect of ovarian hormones on response of the isolated rat uterus to prostaglandins. Prostaglandin Symposium of the Worcester Foundation for Exp. Biol. Eds.: P. W. RAMWELL and J. E. SHAW. New York: Interscience 1968, p. 11—19.
HORTON, E. W.: Action of prostaglandin E_1 on tissues which respond to bradykinin. Nature 200, 892—893 (1963).
— MAIN, I. H. M.: A comparison of the biological activities of four prostaglandins. Brit. J. Pharmac. 21, 182—189 (1963).
— — A comparison of the actions of prostaglandins $F_{2\alpha}$ and E_1 on smooth muscle. Brit. J. Pharmac. 24, 470—476 (1965).
— — The relationship between the chemical structure of prostaglandins and their biological activity. Mem. Soc. Endocr. 14, 29—37 (1966).
— — THOMPSON, C. J.: Effects of prostaglandins on the oviduct, studied in rabbits and ewes. J. Physiol. (Lond.) 180, 514—528 (1965).
KARIM, S. M. M.: Identification of prostaglandins in human amniotic fluid. J. Obstet. Gynaec. Brit. Commonw. 73, 903—908 (1966).
— The identification of prostaglandins in human umbilical cord. Brit. J. Pharmac. 29, 230—237 (1967).
— Appearance of prostaglandin $F_{2\alpha}$ in human blood during labour. Brit. med. J. 4, 618—621 (1968).
— The role of prostaglandin $F_{2\alpha}$ in parturition. In: Prostaglandins, Peptides and Amines. Eds.: P. MANTEGAZZA and E. W. HORTON. London: Academic Press 1969, pp. 65—72.
— Action of prostaglandins in the pregnant woman. Ann. N. Y. Acad. Sci. 180, 483—498 (1971).
— DEVLIN, J.: Prostaglandin content of amniotic fluid during pregnancy and labour. J. Obstet. Gynaec. Brit. Commonw. 74, 230—234 (1967).
— FILSHIE, G. M.: Therapeutic abortion using prostaglandin $F_{2\alpha}$. Lancet 1, 157—159 (1970 a).
— — Use of prostaglandin E_2 for therapeutic abortion. Brit. med. J. 3, 198—200 (1970 b).
— HILLIER, K., TRUSSELL, R. R., PATEL, R. C., TAMUSANGE, S.: Induction of labour with prostaglandin E_2. J. Obstet. Gynaec. Brit. Commonw. 77, 200—210 (1970).
— TRUSSELL, R. R., HILLIER, K., PATEL, R. C.: Induction of labour with prostaglandin $F_2\alpha$. J. Obstet. Gynaec. Brit. Commonw. 76, 769—782 (1969).
— — PATEL, R. C., HILLIER, K.: Response of pregnant human uterus to prostaglandin $F_2\alpha$-induction of labour. Brit. med. J. 4, 621—623 (1968).

KARLSON, S.: The influence of seminal fluid on the motility of the non-pregnant human uterus. Acta obstet. gynec. scand. **38**, 503—521 (1959).

ROTH-BRANDEL, U., BYGDEMAN, M., WIQVIST, N., BERGSTRÖM, S.: Prostaglandins for induction of therapeutic abortion. Lancet. **1**, 190—191 (1970).

SANDBERG, F., INGELMAN-SUNDBERG, A., RYDÉN, G.: The effect of prostaglandin E_1 on the human uterus and the fallopian tubes *in vitro*. Acta obstet. gynec. scand. **42**, 269—278 (1963).

— — — The effect of prostaglandin E_2 and E_3 on the human uterus and fallopian tubes *in vitro*. Acta obstet. gynec. scand. **43**, 95—102 (1964).

— — — The effect of prostaglandin $F_1\alpha$, $B_1\beta$, $F_2\alpha$ and $F_2\beta$, on the human uterus and the fallopian tubes *in vitro*. Acta obstet. gynec. scand. **44**, 585—594 (1965).

— — — JOELSSON, I.: The absorption of tritium-labelled prostaglandin E_1 from the vagina of non-pregnant women. Acta obstet. gynec. scand. **47**, 22—26 (1968).

SULLIVAN, T. J.: Response of the mammalian uterus to prostaglandins under differing hormonal conditions. Brit. J. Pharmac. **26**, 678—685 (1966).

WIQVIST, N., BYGDEMAN, M.: Induction of therapeutic abortion with intravenous prostaglandin F_{2a}. Lancet. **1**, 889 (1970).

— — KWON, S. U., MUKHERJEE, T., ROTH-BRANDEL, U.: Effect of prostaglandin E_1 on the midpregnant human uterus. Amer. J. Obstet. Gynecol. **102**, 327—332 (1968).

VII. Endocrine System

1. Adrenal Cortex

PGE$_2$ (5 µg/ml) mimics the action of corticotrophin upon the rat decapsulated adrenal cortex *in vitro*, stimulating an increased output of corticosterone (FLACK, JESSUP and RAMWELL, 1969). The effects of submaximal doses of PGE$_2$ and corticotrophin are additive

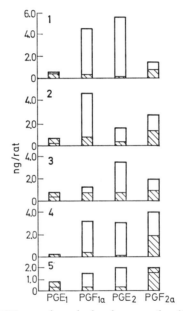

Fig. 1. Effect of ACTH on adrenal gland prostaglandin content *in vivo* in 5 experiments (1—5) using Sprague-Dawley female rats, each weighing 120—140 gm, and dosages of 5 U per kilogram of body weight administered subcutaneously. The values for each experiment were obtained by assaying a pool of adrenal glands taken from 6—8 rats; the values are uncorrected for losses. Recovery of standard prostaglandins was 20—30%. Open bars: unstressed; crosshatched bars: ACTH-treated. Difference between unstressed and ACTH-treated: PGE$_1$, not significant; PGF$_{1\alpha}$, P< 0.02; PGE$_2$, P<0.05; PGF$_{2\alpha}$, P>0.05 < 0.01. (FLACK and RAMWELL, 1970).

and both are similarly reduced by cycloheximide, an inhibitor of protein synthesis. Cyclic AMP and its dibutyryl analogue also stimulate steroidogenesis and it has been suggested that both PGE_2 and corticotrophin exert their effects via a common pathway which includes the adenyl cyclase system (see Chapter XV). FLACK and RAMWELL (1970) observed a *decrease* in adrenal prostaglandin concentrations following corticotrophin administration (Fig. 1). This suggests a further interaction between the two substances on steroidogenesis.

2. Adrenal Medulla

Neither PGE_1 (HORTON, 1963) nor $PGF_{1\alpha}$ (HORTON, unpublished) stimulates adrenaline release from the adrenal medulla of the cat on close-arterial injection. However, the hyperglycaemic response to PGE_1 in the rat is abolished by medullectomy (PAOLETTI, LENTATI and KOROLKIEWICZ, 1967; BERTI, NAIMZADA, LENTATI, USARDI, MANTEGAZZA and PAOLETTI, 1967) and PGE_1 does release catecholamines from the adrenal medulla in the dog (KAYAALP and TÜRKER, 1967). Whether these actions are direct or mediated by the central nervous system has not been established.

3. Corpus Luteum

When $PGF_{2\alpha}$ (1 mg/kg/day) is infused on days 5 and 6 of pseudopregnancy in the rat, either into the right heart or into the lumen of the uterus, ovarian progesterone levels fall, whereas 20α-dihydroprogesterone levels rise, a pattern of change characteristic of luteal regression. On the other hand $PGF_{2\alpha}$ does not affect ovarian steroidogenesis *in vitro*. Subcutaneous $PGF_{2\alpha}$ shortens pseudopregnancy in rats from 17 to 8˙ days as indicated by vaginal smears. Thus $PGF_{2\alpha}$ has a luteolytic action in the rat (PHARRISS and WYNGARDEN, 1969). Since $PGF_{2\alpha}$ is a venoconstrictor, it is suggested that reduction in ovarian blood flow resulting from utero-ovarian venoconstriction may account for its luteolytic action. A similar action has now been observed in several other species. In the rhesus monkey, in which menstruation was induced by $PGF_{2\alpha}$ (30 mg/day) subcutaneously, the corpus luteum was most vulnerable at the later stages of the cycle (KIRTON, PHARRISS and FORBES, 1970). Pseudo-pregnancy in the guinea-pig is shortened by $PGF_{2\alpha}$ (BLATCHLEY and DONOVAN, 1969). Injected subcutaneously in pregnant rabbits (days 4 to 8) $PGF_{2\alpha}$ (5 mg/kg/day) effectively blocked nidation (GUT-KNECHT, CORNETTE and PHARRISS, 1969). Pregnancy in rats is also terminated by this luteolytic action of $PGF_{2\alpha}$ (2—3.2 mg/day, in-jected subcutaneously), though the oral route is not effective. Admi-nistration of a progestogen was capable of maintaining pregnancy in rats which were receiving large doses of $PGF_{2\alpha}$. It is also luteolytic in the sheep (McCRACKEN, GLEW and SCARAMUZZI, 1970).

Neither PGE_1 nor PGE_2 blocks gonadotrophin-stimulated ovarian steroidogenesis in the rabbit, but PGE_2 caused a substantial increase in 20α-dihydroprogesterone (BEDWANI and HORTON, 1968). SPEROFF and RAMWELL (1970) using the bovine corpus luteum *in vitro* showed that PGE_2 and PGE_1 but *not* $PGF_{2\alpha}$ stimulate progesterone synthesis. PGE_2 thus mimicks the action of luteinizing hormone and is ap-proximately half as active as luteinizing hormone on a molar basis. The action of both stimulants is blocked by cycloheximide and both use the same pool of precursors for the synthesis.

Evidence has accumulated that $PGF_{2\alpha}$ but none of the other primary prostaglandins is released from the uterus under conditions associated with the release of the uterine hormone, luteolysin. Thus insertion of a foreign body into the uterus (POYSER, HORTON, THOMP-SON and LOS, 1970, 1971) and the administration of oestrogen on days 4—6 of the oestrous cycle of the guinea-pig (BLATCHLEY, DONO-VAN, POYSER, HORTON, THOMPSON and LOS, 1971) both induce luteolysis and both cause the release of $PGF_{2\alpha}$ from the uterus. This action of $PGF_{2\alpha}$ is therefore likely to prove to be a physiological one (see Chapter XV) though its mechanism of action is still unknown.

4. Thyroid

In canine thyroid slices PGE_1 mimicks the following actions of thyroid stimulating hormone (TSH): increased glucose oxidation, increased colloid droplet formation, increase in ^{131}I release after suppression with thyroxine and increases in cyclic AMP formation and adenyl cyclase activity (ONAYA and SOLOMON, 1970; ZOR, BLOOM, LOWE and FIELD, 1969). The effects of PGE_1 and TSH on glucose oxidation and colloid droplet fromation are additive and both are inhibited by chlorpromazine. Since the thyroid contains high concentrations of prostaglandin-like substances and since both PGE_1 and PGE_2 are released in culture media of medullary carcinoma cells (GRIMLEY, DEFTOS, WEEKS and RABSON, 1969), it seems possible that these compounds are concerned in the secretion of thyroid hormone which is initiated by the trophic hormone, TSH.

5. Islets of Langerhans

PGE_1 increases plasma insulin levels in the mouse (BRESSLER, VARGAS-CORDON and LEBOVITZ, 1968). This is another example of a response which may also involve the adenyl cyclase-cyclic AMP system (see Chapter XV).

References

BEDWANI, J. R., HORTON, E. W.: The effects of prostaglandins E_1 and E_2 on ovarian steroidogenesis. Life Sci. 7, 389—393 (1968).

BERTI, F., NAIMZADA, M. K., LENTATI, R., USARDI, M. M., MANTEGAZZA, P., PAOLETTI, R.: Relations between some in vitro and in vivo effects of prostaglandin E_1. Prog. biochem. Pharmac. 3, 110—121 (1967).

BLATCHLEY, F. R., DONOVAN, B. T.: Luteolytic effect of prostaglandin in the guinea-pig. Nature 221, 1065—1066 (1969).

— — POYSER, N. L., HORTON, E. W., THOMPSON, C. J., Los, M.: Identification of prostaglandin $F_{2\alpha}$ in the utero-ovarian blood of guinea-pig following oestrogen treatment. Nature 230, 243—244 (1971).

BRESSLER, R., VARGAS-CORDON, M., LEBOVITZ, H. E.: Tranylcypromine: a potent insulin secretagogue and hypoglycemic agent. Diabetes 17, 617—624 (1968).

FLACK, J. D., JESSUP, R., RAMWELL, P. W.: Prostaglandin stimulation of rat corticosteroidogenesis. Science, N. Y. 163, 691—692 (1969).

— RAMWELL, P. W.: Cited by P. W. RAMWELL and J. E. SHAW. Recent Prog. Horm. Res. 26, 150 (1970).

GRIMLEY, P. M., DEFTOS, L. J., WEEKS, J. R., RABSON, A. S.: Growth *in vitro* and ultrastructure of cells from a medullary carcinoma of the human thyroid gland: Transformation by Simian virus 40 and evidence of thyrocalcitonin and prostaglandins. J. natn. Cancer Inst. 42, 663—680 (1969).

GUTKNECHT, G. D., CORNETTE, J. C., PHARRISS, B. B.: Antifertility properties of prostaglandin $F_{2\alpha}$. Biol. Reprod. 1, 367—371 (1969).

HORTON, E. W.: Action of prostaglandin E_1 on tissues which respond to bradykinin. Nature 200, 892—893 (1963).

KANEKO, T., ZOR, U., FIELD, J. B.: Thyroid-stimulating hormone and prostaglandin E_1 stimulation of cyclic 3′,5′-adenosine monophosphate in thyroid slices. Science, N. Y. 163, 1062—1063 (1969).

KAYAALP, S. O., TÜRKER, R. K.: Release of catecholamines from the adrenal medulla by prostaglandin E_1. Eur. J. Pharmac. 2, 175—180 (1967).

KIRTON, K. T., PHARRISS, B. B., FORBES, A. D.: Luteolytic effects of prostaglandin $F_{2\alpha}$ in primates. Proc. Soc. exp. Biol. Med. 133, 314—316 (1970).

McCRACKEN, J. A., GLEW, M. E., SCARAMUZZI, R. J.: Corpus luteum regression induced by prostaglandin $F_{2\alpha}$. J. clin. Endocr. Metab. 30, 544—546 (1970).

ONAYA, T., SOLOMON, D. H.: Stimulation by prostaglandin E_1 of endocytosis and glucose oxidation in canine thyroid slices. Endocrinology 86, 423—426 (1970).

PAOLETTI, R., LENTATI, R. L., KOROLKIEWICZ, Z.: Pharmacological investigations on the prostaglandin E_1 effect on lipolysis. Nobel Symposium 2, Prostaglandins. Ed.: S. BERGSTRÖM and B. SAMUELSSON. Stockholm: Almqvist and Wiksell 1967, pp. 147—159.

PHARRISS, B. B., WYNGARDEN, L. J.: The effect of prostaglandin $F_{2\alpha}$ on the progestogen content of ovaries from pseudopregnant rats. Proc. Soc. exp. Biol. Med. 130, 92—94 (1969).

POYSER, N. L., HORTON, E. W., THOMPSON, C. J., LOS, M.: Identification of prostaglandin $F_{2\alpha}$ released by distension of guinea pig uterus in vitro. J. Endocrinol. 48, xliii (1970).

— — — — Identification of prostaglandin $F_{2\alpha}$ released by distension of the guinea-pig uterus in vitro. Nature, 230, 526—528 (1971).

SPEROFF, L., RAMWELL, P. W.: Prostaglandin stimulation of in vitro progesterone synthesis. J. clin. Endocr. Metab. 30, 345—350 (1970).

ZOR, U., BLOOM, G., LOWE, I. P., FIELD, J. B.: Effects of theophylline, prostaglandin E_1 and adrenergic blocking agents on TSH stimulation of thyroid intermediary metabolism. Endocrinology 84, 1082—1088 (1969).

VIII. Adipose Tissue

1. In vitro Experiments

The anti-lipolytic action of PGE₁ was first reported by STEIN-BERG, VAUGHAN, NESTEL and BERGSTRÖM in 1963. They observed that PGE₁ (20—100 ng/ml) inhibits the fat mobilising action of adrenaline, noradrenaline, corticotrophin, glucagon and thyroid stimulating hormone on rat epididymal adipose tissue *in vitro* (Fig. 1). They suggested that the action of PGE₁ was at a point in the biochemical pathway common to the action of all these hormones. Other prostaglandins are less potent than PGE₁ on lipolysis (STEINBERG, VAUGHAN, NESTEL, STRAND and BERGSTRÖM, 1964).

PGE₁ inhibits basal lipolysis (STEINBERG et al., 1963, 1964), lipolysis in alloxan-diabetic rats (STOCK and WESTERMANN, 1966)

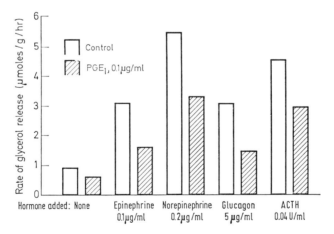

Fig. 1. PGE₁ inhibition of glycerol release from rat epididymal fat pads in vitro. Six pairs of tissues in each hormone study were incubated for 1 hour in 3 ml of Krebs' bicarbonate solution containing bovine serum albumin, 30 mg/ml, in 95⁰/o oxygen, 5⁰/o carbon dioxide. Hormones were added to both flasks; PGE₁ (0.1 µg/ml) to only one. The data on tissues not exposed to hormone represent 16 pairs of tissues (STEINBERG et al., 1964)

and also lipolysis induced by theophylline (STEINBERG and VAUGHAN, 1967; PAOLETTI, LENTATI and KOROLKIEWICZ, 1967; MÜHLBACHOVÁ, SÓLYOM and PUGLISI, 1967). Lipolysis induced by fasting is not inhibited by PGE_1. Nicotinic acid, on the other hand, inhibits lipolysis in fat from both fasted and fed animals (BERGSTRÖM and CARLSON, 1965 a; STOCK and WESTERMANN, 1966).

The inhibitory action of PGE_1 on basal, theophylline-induced and hormone-induced lipolysis has also been observed in isolated white and brown fat cells incubated in the presence of glucose (FAIN, 1967, 1968). This established that PGE_1 acts directly on the adipocyte. There is some disagreement about the action of PGE_1 on brown fat cells but it is likely that plasma levels of free fatty acids and glycerol in the intact animal reflect mobilisation of white rather than brown fat (BIZZI, CODEGONI, LIETTI and GARATTINI, 1968).

Noradrenaline and corticotrophin activate the enzyme adenyl cyclase in adipose tissue, thus promoting the conversion of adenosine triphosphate (ATP) to cyclic $3'5'$-adenosine monophosphate (cyclic AMP). The degradation of cyclic AMP to adenosine monophosphate (AMP) is catalysed by the enzyme, phosphodiesterase, which can be blocked by xanthines like theophylline or caffeine. Accumulation of cyclic AMP activates lipolytic enzymes. Thus hormones by activating adenyl cyclase and theophylline by blocking the inactivation of cyclic AMP both cause lipolysis. PGE_1 blocks lipolysis induced by either of these mechanisms but it does not block lipolysis induced by cyclic AMP or its dibutyryl analogue. It therefore appears that PGE_1 acts by inhibition of adenyl cyclase and thus the synthesis of cyclic AMP (STEINBERG and VAUGHAN, 1967). Evidence in support of this hypothesis (see also Chapter XV) has been obtained by estimation of cyclic AMP levels in adipose tissue. The increase in concentration of cyclic AMP in isolated fat cells in response to noradrenaline can be prevented by PGE_1 (BUTCHER, PIKE and SUTHERLAND, 1967). One suggestion is that PGE_1 interferes with the binding of ATP to adenyl cyclase (STOCK, AULICH and WESTERMANN, 1968). Interaction of prostaglandins with the adenyl cyclase system are discussed in Chapter XV. The possibility that PGE_1 causes increased resynthesis of triglyceride rather than reduced breakdown seems to have been excluded (SÓLYOM, PUGLISI and MÜHLBACHOVÁ, 1967).

The mechanism of action of PGE_1 on adipose tissue is, however, more complex. PGE_1 in the absence of a lipolytic hormone *increases* the concentration of cyclic AMP in adipose tissue, though not in isolated fat cells, yet PGE_1 does not stimulate lipolysis in the isolated preparation. The possibility that other cells in adipose tissue contribute to the increased cyclic AMP levels in response to PGE_1 has been suggested (BUTCHER and BAIRD, 1968).

In rabbit perirenal adipose tissue *in vitro*, noradrenaline stimulates lipolysis which is antagonised by PGE_1 (BOBERG, MICHELI and RAMMER, 1970).

2. In vivo Experiments

In anaesthetised and unanaesthetised dogs injections or infusions of PGE_1 in low dosage (0.2 µg/kg/min) raise plasma free fatty acid levels. This lipolytic action can be abolished by blockade of the autonomic ganglia and appears to be an indirect action upon adipose tissue (BERGSTRÖM, CARLSON and ORÖ, 1966 a). In higher doses PGE_1 has an antilipolytic effect (Fig. 2) (BERGSTRÖM, CARLSON and ORÖ, 1964, 1966 b), reversing plasma free fatty acid levels raised by nor-

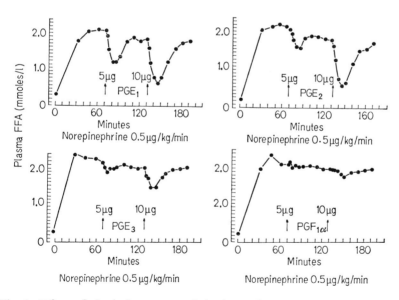

Fig. 2. Effect of single intravenous injections of PGE_1, PGE_2, PGE_3 and $PGF_{1\alpha}$ on the plasma levels of free fatty acids in anaesthetised dogs during infusions of noradrenaline (BERGSTRÖM et al., 1964)

adrenaline infusion and this appears to be a direct action upon adipose tissue, like the response observed *in vitro* in rats.

However, the results obtained with dog omental adipose tissue *in vitro* differ in the opposite direction (CARLSON and MICHELI, 1970). PGE_1 (0.1 µg/ml) *in vitro* inhibited basal lipolysis whereas higher doses stimulated lipolysis. The lipolytic action *in vitro* was not abolished by blockade of adrenaline beta receptors. A further complication is the finding that in only 5 out of 10 dogs did PGE_1

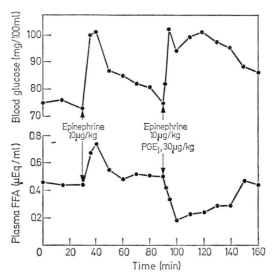

Fig. 3. Inhibition of the lipid-mobilising action of adrenaline in a conscious dog by simultaneous intravenous injection of PGE_1 (STEINBERG and PITT-MANN, 1966)

inhibit noradrenaline-induced glycerol release *in vitro*. These puzzling discrepancies raise the possibility that adipose tissue from different sites and in different species differs in the mechanism of its response to PGE_1.

PGA_1 (20 µg/kg) which has a depressor effect like PGE_1 produced a slight rise in fatty acid levels in the dog (STEINBERG and PITTMAN, 1966). PGE_1 has no effect on the hyperglycaemia induced by adrenaline injection (Fig. 3) STEINBERG and PITTMAN, 1966; BERGSTRÖM et al., 1966 a).

PGE$_1$ (5.6 µg/kg/min) infused intravenously in rats inhibits basal lipolysis in fed but not fasted animals, and inhibits lipolysis induced by corticotrophin, noradrenaline and exposure to cold (BERTI, LENTATI, USARDI and PAOLETTI, 1967). KUPIECKI (1967) differing from other workers observed inhibition of basal lipolysis in fasted rats with PGE$_1$ both *in vivo* and *in vitro*. A difference in strain is one explanation for this discrepancy.

3. Human Adipose Tissue

Human subcutaneous adipose tissue in vitro releases free fatty acids and glycerol in response to noradrenaline. This lipolytic action is inhibited by PGE$_1$ (1 µg/ml) (BERGSTRÖM and CARLSON, 1965 b). A similar inhibitory effect has been observed with PGE$_1$ on human omental fat and there is no obvious difference in response with sex or age (CARLSON and HALLBERG, 1968; MICHELI, CARLSON and HALLBERG, 1969).

In contrast to this antilipolytic action, PGE$_1$ (32—580 ng/kg/min) infused intravenously in human volunteers raised plasma free fatty acid and glycerol levels (BERGSTRÖM, CARLSON, EKELUND and ORÖ, 1965; CARLSON, 1967). By extrapolation from results obtained in dogs it seems probable that this lipolytic effect is an indirect one dependent upon sympathetic pathways and that a direct antilipolytic action on adipose tissue would be observed if larger amounts were infused. Such larger doses are prohibited by the side effects encountered with PGE$_1$ (CARLSON, 1967).

4. Obesity

The discovery of the antilipolytic action of PGE$_1$ has raised the question of the possible implication of this substance in obesity. It is known, for example, that in certain forms of experimental obesity, mobilisation of fat in response to normal stimuli is impaired. There is consequently great interest in the observation by HAESSLER and CRAWFORD (1966) that in the adipose tissue of rats made obese by electrolytic lesions in the hypothalamus, there is an inhibitor of lipolysis. This substance is a polar lipid with some of the biological actions of a prostaglandin, in particular it is a powerful inhibitor of noradrenaline-stimulated lipolysis.

If an increased production of antilipolytic prostaglandins in fat from obese animals can be confirmed, the possibility of reversing this with prostaglandin antagonists may hold out hope for the treatment of some types of human obesity.

References

BERGSTRÖM, S., CARLSON, L. A.: Influence of the nutritional state on the inhibition of lipolysis in adipose tissue by prostaglandin E_1 and nicotinic acid. Acta physiol. scand. **65**, 383—384 (1965 a).

— — Inhibitory action of prostaglandin E_1 on the mobilization of free fatty acids and glycerol from human adipose tissue *in vitro*. Acta physiol. scand. **63**, 195—196 (1965 b).

— — EKELUND, L. G., ORÖ, L.: Cardiovascular and metabolic response to infusions of prostaglandin E_1 and to simultaneous infusions of nor-adrenaline and prostaglandin E_1 in man. Acta physiol. scand. **64**, 332—339 (1965).

— — ORÖ, L.: Effect of prostaglandins on catecholamine induced changes in the free fatty acids of plasma and in blood pressure in the dog. Acta physiol. scand. **60**, 170—180 (1964).

— — — Effect of different doses of prostaglandin E_1 on free fatty acids of plasma, blood glucose and heart rate in the non-anaesthetised dog. Acta physiol. scand. **67**, 185—193 (1966 a).

— — — Effect of prostaglandin E_1 on plasma free fatty acids and blood glucose in the dog. Acta physiol. scand. **67**, 141—151 (1966 b).

BERTI, F., LENTATI, R., USARDI, M. M., PAOLETTI, R.: The effect of pro-staglandin E_1 on free fatty acid mobilization and transport. Proto-plasma. **63**, 143—146 (1967).

BIZZI, A., CODEGONI, A. M., LIETTI, A., GARATTINI, S.: Different responses of white and brown adipose tissue to drugs affecting lipolysis. Biochem. Pharmac. **17**, 2407—2412 (1968).

BOBERG, J., MICHELI, H., RAMMER, L.: Effect of nicotinic acid on ACTH and noradrenaline stimulated lipolysis in the rabbit. II. *In vitro* studies including comparison with prostaglandin E_1. Acta physiol. scand. **79**, 299—304 (1970).

BUTCHER, R. W., BAIRD, C. E.: Effects of prostaglandins on adenosine 3',5'-monophosphate levels in fat and other tissues. J. biol. Chem. **243**, 1713—1717 (1968).

BUTCHER, R. W., PIKE, J. E., SUTHERLAND, E. W.: The effect of prostaglandin E_1 on adenosine 3',5'-monophosphate levels in adipose tissue. Nobel Symposium 2, Prostaglandins. Ed.: S. BERGSTRÖM and B. SAMUELSSON. Stockholm: Almqvist and Wiksell 1967, pp. 133—138.

CARLSON, L. A.: Metabolic and cardio-vascular effects *in vivo* of prostaglandins. Nobel Symposium 2, Prostaglandins. Ed.: S. BERGSTRÖM and B. SAMUELSSON. Stockholm: Almqvist and Wiksell 1967, pp. 123—132.

— HALLBERG, D.: Basal lipolysis and effects of norepinephrine and prostaglandin E_1 on lipolysis in human subcutaneous and omental adipose tissue. J. Lab. clin. Med. **71**, 368—377 (1968).

— MICHELI, H.: Some characteristics of lipolysis in dog adipose tissue. Effects of noradrenaline, prostaglandin E_1 and nicotinic acid. Acta physiol. scand. **79**, 145—152 (1970).

FAIN, J. N.: Adrenergic blockade of hormone-induced lipolysis in isolated fat cells. Ann. N. Y. Acad. Sci. **139**, 879—890 (1967).

— Antilipolytic effect of prostaglandin E_1 on free fat cells. Prostaglandin Symposium of the Worcester Foundation for Exp. Biol. Ed.: P. W. RAMWELL and J. E. SHAW. New York: Interscience 1968, pp. 67—77.

HAESSLER, H. A., CRAWFORD, J. D.: Lipolysis in homogenates of adipose tissue: An inhibitor found in fat from obese rats. Science, N. Y. **154**, 909—910 (1966).

KUPIECKI, F. P.: Effects of prostaglandin E_1 on lipolysis and plasma free fatty acids in the fasted rat. J. Lipid Res. **8**, 577—580 (1967).

MICHELI, H., CARLSON, L. A., HALLBERG, D.: Comparison of lipolysis in human subcutaneous and omental adipose tissue with regard to effects of noradrenaline, theophylline, prostaglandin E_1 and age. Acta chir. scand. **135**, 663—670 (1969).

MÜHLBACHOVÁ, E., SÓLYOM, A., PUGLISI, L.: Investigations on the mechanism of the prostaglandin E_1 antagonism to norepinephrine and theophylline-induced lipolysis. Eur. J. Pharmac. **1**, 321—325 (1967).

PAOLETTI, R., LENTATI, R. L., KOROLKIEWICZ, Z.: Pharmacological investigations on the prostaglandin E_1 effect on lipolysis. Nobel Symposium 2, Prostaglandins. Ed.: S. BERGSTRÖM and B. SAMUELSSON. Stockholm: Almqvist and Wiksell 1967, pp. 147—159.

SÓLYOM, A., PUGLISI, L., MÜHLBACHOVÁ, E.: Effect of *in vitro* theophylline and prostaglandin E_1 on free fatty acid release and on triglyceride synthesis in rat adipose tissue. Biochem. Pharmac. **16**, 521—525 (1967).

STEINBERG, D., PITTMAN, R.: Depression of plasma FFA levels in unanaesthetised dogs by single intravenous doses of prostaglandin E_1. Proc. Soc. exp. Biol. Med. **123**, 192—196 (1966).

— VAUGHAN, M.: *In vitro* and *in vivo* effects of prostaglandins on free fatty acid metabolism. Nobel Symposium 2, Prostaglandins. Ed.: S. BERGSTRÖM and B. SAMUELSSON. Stockholm: Almqvist and Wiksell 1967, pp. 109—121.

STEINBERG, D., VAUGHAN, M., NESTEL, P., BERGSTRÖM, S.: Effects of prostaglandin E opposing those of catecholamines on blood pressure and on triglyceride breakdown in adipose tissue. Biochem. Pharmac. **12**, 764—766 (1963).

— — — STRAND, O., BERGSTRÖM, S.: Effects of the prostaglandins on hormone-induced mobilization of free fatty acids. J. clin. Invest. **43**, 1533—1540 (1964).

STOCK, K., AULICH, A., WESTERMANN, E.: Studies on the mechanism of antilipolytic action of prostaglandin E_1. Life Sci. **7**, 113—124 (1968).

— WESTERMANN, E.: Hemmung der Lipolyse durch α- und β-Sympathicolytica, Nicotin-saure und Prostaglandin E_1. Arch. Pharmak. exp. Path. **254**, 334—354 (1966).

IX. The Nervous System

1. Central Nervous System

Amongst the most remarkable actions of the prostaglandins are their effects on the central nervous system. Description of these is not easy since different prostaglandins have different effects and since the response observed varies not only with the species but upon the presence or absence of higher centres or of anaesthetic.

The original pharmacological observations were made on the young chick and the unanaesthetised cat with a chronically indwelling cannula in one lateral cerebral ventricle (HORTON, 1964). The choice of these preparations was determined by the small amounts of prostaglandins available in 1963. The young chick is believed to lack a blood brain barrier and it is known to be sensitive to small quantities of various centrally-acting drugs. The injection of drugs into the lateral ventricle of an unanaesthetised cat is a convenient way of circumventing the blood brain barrier and of bringing small quantities of a drug into contact with several structures which impinge upon the ventricular system, all in the absence of anaesthetic. Both test systems indicated that PGE compounds have effects on central pathways. These are described below.

Actions on a proportion of central neurones was subsequently demonstrated by micro-iontophoretic application to single cells in the medulla (AVANZINO, BRADLEY and WOLSTENCROFT, 1966). This direct evidence was valuable in establishing that central actions of prostaglandins do not depend upon vascular changes.

a) Sedative Actions of PGE in the Unanaesthetised Chick

When injected into intact unanaesthetised animals PGE compounds have sedative-tranquillizer actions. Thus in the chick, PGE_1, PGE_2 and PGE_3 injected intravenously cause profound sedation (HORTON, 1964). After the injection the chick lies on its side and with

Fig. 1. (a) Chick 39 g photographed 1 minute after an intravenous injection of 2 µg of PGE₁. (b) Same chick 30 minutes later. (HORTON and MAIN, 1965)

the higher doses (50 to 200 µg/kg PGE₁) the righting reflex is lost (Fig. 1 a). There is little spontaneous activity apart from a few short-lasting convulsive movements of the legs and wings soon after the injection. The duration of the sedation varies from 5 min at the 10 µg/kg dose level to 75 min at 200 µg/kg. As sedation subsides, the chick slowly resumes its normal posture. The righting reflex is first restored, the chick remaining in a squatting position with its eyes closed (Fig. 1 b). Gradually the chick resumes the standing posture and finally it begins to chirp and move about the cage. During the period of sedation the chicks are invariably silent even when pressure is applied to a toe, although this produces immediate withdrawal of the limb. There is no loss of the corneal reflex and, when the standing posture has been regained, there is no abnormality of gait or lack of co-ordination.

b) Sedative-tranquillizer Actions of PGE₁ in the Mouse

The sedative action of PGE₁ has been investigated further in the mouse (HOLMES and HORTON, 1968 a). On subcutaneous or intravenous injection PGE₁ (1 mg/kg) causes diminished general motor activity (lack of exploratory activity and washing movements) and ptosis (Fig. 2) lasting 30—60 minutes; the animals appear completely

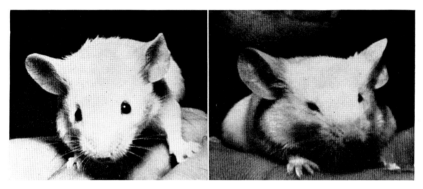

Fig. 2. Ptosis induced by PGE₁. (a) mouse injected subcutaneously with saline (10 ml/kg), (b) mouse injected subcutaneously with PGE₁ (1 mg/kg). Photographs taken 10 minutes after injection. (HOLMES and HORTON, 1969 a)

normal 90 minutes after injection. Control mice dosed with saline solution under the same environmental conditions show considerable exploratory activity and washing movements during the first 30 minutes after injection. The PGE₁ treated mice produce moist faeces during the first 30 minutes after injection but there is no increase in frequency of defaecation. No sign of ataxia, lacrimation, excess urination, piloerection, convulsions, tremor, salivation, cyanosis or vocalisation is observed during the 2-hour observation period; the arousal to auditory and tactile stimuli appears to be unimpaired; the respiratory rate is not significantly altered; the righting, pinna, corneal and place reflexes are normal and there is no change in the ability of mice to stay on an inclined wire screen or climb a wooden pole. After intravenous injection vasodilatation in the tail is noted.

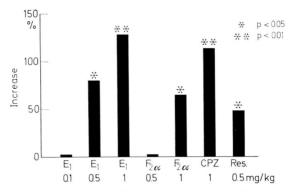

Fig. 3. Potentiation of hexobarbitone sleeping time in mice. Groups of 10 mice (18—22 g), 5 of each sex in room temperature of 28—30° C. Drugs injected subcutaneously followed after 5 minutes by an intraperitoneal injection of hexobarbitone (85 mg/kg). Time from loss of righting reflex to recovery of righting reflex measured. Ordinate: percentage increase in sleeping time over control animals (saline treated). Abscissa: $E_1 = PGE_1$; $F_{2\alpha} = PGF_{2\alpha}$; CPZ = chlorpromazine; Res. = reserpine. $** = P < 0.01$; $* = P < 0.05$. (HOLMES and HORTON, 1968 a)

PGE_1 (0.5 mg/kg) causes a significant ($P < 0.05$) potentiation of hexobarbitone sodium sleeping time in mice and a dose of 1 mg/kg produces a highly significant potentiation ($P < 0.01$). $PGF_{2\alpha}$ also potentiates hexobarbitone sleeping time but is less active than PGE_1; 1 mg/kg produced a significant potentiation ($P < 0.05$) but 0.5 mg/kg did not ($P > 0.05$) (Fig. 3) (HOLMES and HORTON, 1968 a). PGE_1 is about equiactive with chlorpromazine in this test but the time course of action of the two compounds is different.

c) Anticonvulsant Actions of PGE_1

PGE_1 exhibited some protection against leptazol (100 mg/kg) induced convulsions at a dose of 0.5 mg/kg and exhibited almost complete protection at the 1 mg/kg level (Fig. 4). The protection appeared to be of short duration since it was more marked in the first 10 min than in the first 30 min after challenge with leptazol.

In a second study carried out on a different strain of mice it was found that PGE₁, 1 mg/kg subcutaneously, provided no protection against leptazol induced clonic and tonic convulsions or death. In this strain PGE₁, 10 mg/kg, gave complete protection against leptazol induced tonic extensor convulsions and death but no protection

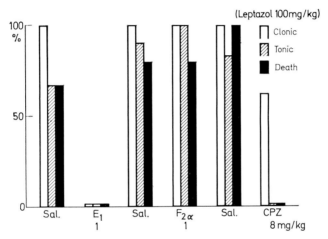

Fig. 4. Leptazol convulsions in mice. PGE₁ (6 mice) and PGF₂ₐ (10 mice) injected subcutaneously 5 minutes before an intraperitoneal injection of leptazol (100 mg/kg). Chlorpromazine (6 mice) injected subcutaneously 30 minutes before the leptazol. Ordinate: percentage of mice convulsing (clonic or tonic) or dying within 10 minutes of challenge. Abscissa: Sal. = saline; E₁ = PGE₁; F₂ₐ = PGF₂ₐ; CPZ = chlorpromazine.
(HOLMES and HORTON, 1968 a)

against clonic convulsions. Thus it would appear that there may be strain variation in the anti-convulsant potency of PGE₁.

Chlorpromazine hydrochloride, 8 mg/kg, protected the mice against tonic convulsions and death but not against clonic convulsions. Reserpine, 5 mg/kg, provided little protection against the effects of leptazol.

The short duration of the anti-convulsant actions of PGE₁ is also shown in its antagonism of tonic extensor seizures due to maximal electro-shock; some protection is observed at 10 minutes with 0.2 mg/

kg of PGE$_1$, whereas 1 mg/kg is inactive at 60 minutes. Doses of 0.2—1 mg/kg give complete protection against the lethal actions of maximal electro-shock. Chlorpromazine hydrochloride, 8 mg/kg, provides some protection against death but not against tonic extensor seizures. Phenobarbitone sodium 50 mg/kg, provides complete protection against both tonic extensor seizures and death (Fig. 5).

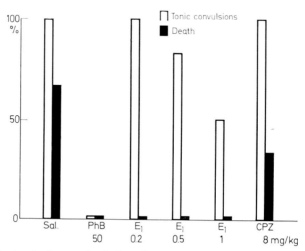

Fig. 5. Maximal electro-shock in mice. Groups of 6 mice. Sodium phenobarbitone (PhB) and PGE$_1$ injected subcutaneously 10 minutes before challenge and chlorpromazine injected subcutaneously 30 minutes before challenge with shocks of 80 volts and 0.2 seconds duration applied with spring-loaded electrodes placed in the ears. Ordinate as Fig. 4. (HOLMES and HORTON, 1968 a)

PGE$_1$ (1 mg/kg) provides some protection against both convulsions and death due to strychnine (1 mg/kg) but is inactive against strychnine (2 mg/kg). PGE$_1$ (1 mg/kg) was inactive against both the convulsant and lethal effects of picrotoxin (10 mg/kg) which was the threshold dose for lethal action within 30 min of challenge. DURU and TÜRKER (1969) report antagonism of strychnine convulsions following much lower doses of PGE$_1$ (20—30 µg/kg) in mice.

d) Other Actions of PGE$_1$ in the Mouse

PGE$_1$, 1 mg/kg, exhibited slight analgesic action as judged by the acetylcholine-induced writhing test (Fig. 6) (HOLMES and HORTON,

1968 a). At this dose level PGE₁ causes a degree of ataxia (or motor discoordination) as measured on the rotarod. A 50% fall out occurred 10 min after dosing but the effect was wearing off 30 and 60 minutes after dosing. Chlorpromazine hydrochloride, 4 mg/kg, and reserpine, 2 mg/kg, were also shown to produce considerable fall out.

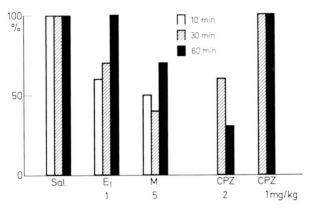

Fig. 6. Abdominal writing test. Acetylcholine chloride 5 mg/kg injected intraperitoneally was used as the nociceptive agent (COLLIER, HAMMOND, HORWOOD-BARRETT and SCHNEIDER, 1964). Groups of 10 mice. Ordinate: percentage of mice showing positive response to acetylcholine at 10, 30 and 60 minutes after treatment. Abscissa: Sal. = saline; E₁ = PGE₁ (1 mg/kg); M = morphine sulphate (5 mg/kg); CPZ = chlorpromazine hydrochloride (2 mg/kg and 1 mg/kg). (HOLMES and HORTON, unpublished)

The acetylcholine writhing or "abdominal constriction" test is known to produce large numbers of false positive reactions when it is employed as a test for analgesia. COLLIER (1964) suggests that if tests for muscular discoordination such as the rotarod are carried out at the same time as the abdominal constriction test many of these false positives can be eliminated by disregarding compounds which give a $\frac{\text{dose preventing constriction}}{\text{dose causing discoordination}}$ ratio greater than 0.5. On this basis PGE₁ cannot be considered an analgesic. No sign of analgesic actions was observed when the tail clip test (BIANCHI and FRANCESCHINI, 1954) was employed.

Ptosis in mice caused by PGE₁ (1 mg/kg) was prevented by pretreatment with imipramine injected 30 minutes before, but iproniazid pre-treatment was ineffective, neither did iproniazid reverse the general motor depression.

PGE$_1$ (1 mg/kg) pre-treatment did not prevent the morphine-induced Straub tail and circling, or the tremor and salivation induced by oxotremorine in mice.

In two experiments mice killed 20 min after the subcutaneous administration of PGE$_1$, 1 mg/kg, showed a 14% reduction in brain catecholamine levels (compared with those of control groups dosed with saline). In a third experiment the animals killed 40 min after the administration of the same dose of PGE$_1$ showed only a 9% depletion in brain catecholamine levels, whereas animals killed 3 hr after the administration of reserpine, 2.5 mg/kg, showed a 67% depletion (HOLMES, 1968). It would be of interest to know whether larger doses of PGE$_1$ would produce less marginal changes in catecholamine levels.

e) Actions of PGE in the Unanaesthetised Cat

When PGE$_1$ is injected intravenously in the cat in doses up to 30 µg/kg there is some diminution in spontaneous activity but minimal sedation. Higher doses have not been tested. On the other hand when injected through a chronically implanted cannula into a lateral ventricle of an unanaesthetised cat, PGE$_1$ (3—20 µg/kg) after a latent period of 5 to 20 minutes causes signs of sedation and stupor. The following account is taken from HORTON (1964):

"Spontaneous movement decreased and the cat would sit in a corner of its cage or, if allowed its freedom, would seek out a dark recess usually under a bench, where it would continue to sit for hours, if undisturbed. The cat assumed a characteristic posture with head forward and slightly lowered; the eyes were closed. The cat showed little interest in its surroundings, it did not resent being picked up and it showed no signs of affection. When taken up and set free, it tended to retire quickly to its former sheltered position. Its movements were not impaired but rapid and fully co-ordinated.

The cat failed to respond to a sudden loud noise or to a bright flash of light. On the other hand, when pressure was applied to a food-pad the limb was rapidly withdrawn but there was no vocalization, indeed the cats were invariably silent. To a few stimuli, there was a sluggish response; for example, on the introduction of another cat into the room, the eyes slowly opened, the ears pricked up and the head moved in the appropriate direction.

Sometimes there were definite signs of catatonia. This was a late feature which occurred after a latency of at least 40 min and developed gradually. When fully developed, the cat could be placed across the rungs of an inverted stool and would remain in such an unnatural position without moving for periods up to 90 min. In contrast, uninjected cats could not be induced to adopt such a position at all. The catatonic signs lasted up to 4 hr, sedation and stupor up to 24 hr, and even at 48 hr there was sometimes reduced spontaneous activity. The threshold dose (of PGE_1) which produced sedation and stupor was approximately 3 μg/kg and the effect lasted 4 to 8 hr.

Another effect of the injection of PGE_1 was moderate dilatation of the pupils lasting 3 to 4 hr; the pupillary reflexes were however normal. There was no evidence of any loss of function of any cranial or spinal nerves. There was no obvious change in respiratory or cardiac rates; no salivation, lachrymation, vomiting, defaecation or micturition occurred. There was no hyperphagia, indeed cats which had been starved for 24 hr before the injection showed no interest in food during stupor due to PGE_1. The injection caused no scratching or other movements.

In one cat PGE_2 (12 μg/kg) and in another cat PGE_3 (12 μg/kg) were injected intraventricularly. Sedation, stupor and catatonia developed and the effects resembled those seen following an intraventricular injection of 7 μg/kg of PGE_1.

In control experiments intraventricular injections of neutral 0.9% saline were followed by transient slight diminution in spontaneous activity."

During the catatonic phase which is maximal between 1 and 4 hours after injection, when the animal is placed on its side a forelimb can be flexed so that the paw rests behind the neck or upper thorax — the position being retained for several minutes without interference by the experimenter.

The effects following an intraventricular injection of PGE compounds (Fig. 7 a) resemble the late effects of intraventricular injection of physostigmine, dyflos, acetylcholine or bulbocapnine seen by FELDBERG and SHERWOOD (1954, 1955). These authors describe the posture of the cat after intraventricular physostigmine as sitting "hunched up without movement; its eyes half shut or shut and its head slightly

a

b

Fig. 7. (a) Cat 4.6 kg photographed 1 hour after an injection into a lateral cerebral ventricle of PGE_1, 20 µg. (b) Same cat photographed 1 hour after an injection into a lateral cerebral ventricle of $PGF_{2\alpha}$, 15 µg. There was a 13 day interval between the two injections. (HORTON and MAIN, 1965)

inclined forward"; this description could be used equally for cats injected with PGE. The catatonic stupor produced by the four drugs used by FELDBERG and SHERWOOD, however, was preceded by signs of pronounced excitation — vigorous scratching, licking and washing movements followed by reflex hyperexcitability and tremor. None of these excitatory effects was observed with the prostaglandins.

f) Actions of PGF$_{2\alpha}$ on Unanaesthetised Animals

In contrast to these effects with PGE compounds, PGF$_{2\alpha}$ has weak sedative actions. In the cat intraventricular injection of PGF$_{2\alpha}$ in doses equal to and higher than those of PGE$_1$ required to produce stupor and catatonia, had no obvious effect. The cats showed no signs of sedation but remained alert and interested in their surroundings (Fig. 7b). There was no stupor and no catatonia (HORTON and MAIN, 1965).

In mice, subcutaneous injection of PGF$_{2\alpha}$ (1 mg/kg) caused no obvious change in spontaneous activity and there was no ptosis. On the other hand PGF$_{2\alpha}$ (1 mg/kg) did potentiate hexobarbitone sleeping time (Fig. 3). Unlike PGE$_1$ it did not antagonise leptazol convulsions but like amphetamine it potentiated the effects of subthreshold convulsant doses of leptazol (Fig. 8).

Neither PGF$_{1\alpha}$ or PGF$_{2\alpha}$ causes sedation in chicks. However PGF$_{2\alpha}$ (25—450 µg/kg) injected intravenously in young chicks causes an immediate and extreme extension of the limbs with some dorsiflexion of the neck (Fig. 9). These effects last about 2 to 10 minutes. During this period the chick is unable to stand or to right itself when placed on its back. The chick's legs are unable to support the weight of its body, but the legs can be moved when support is provided. There is no evidence of sedation, no effect on respiration and the eyes are open throughout. When the chick regains the standing position the legs tends to separate until gradually the chick's abdomen comes to rest on the bench with its legs in full abduction. This tendency for the legs to abduct lasts for up to 30 minutes from the time of injection. With the lower dose range (5—25 µg/kg) abduction of the legs is the only effect seen with PGF$_{2\alpha}$ (HORTON and MAIN, 1965).

Extension of the limbs on intravenous injection of PGF$_{2\alpha}$ is also observed in anaesthetised chicks. In two chicks lightly anaesthetised with urethane (1.25 g/kg) in which the righting reflex was absent, the

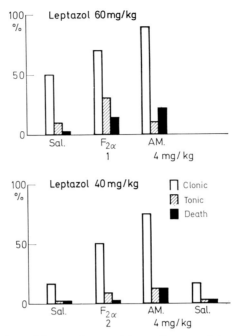

Fig. 8. Potentiation of leptazol convulsions. Groups of 6 to 12 mice. PGF$_{2\alpha}$ injected subcutaneously 5 minutes before challenge. Amphetamine sulphate (AM) injected subcutaneously 30 minutes before challenge with leptazol 60 mg/kg (upper figure) and 40 mg/kg (lower figure) injected intraperitoneally. Ordinate: percentage of animals convulsing (clonic and tonic) and dying within 10 minutes of challenge. (HOLMES and HORTON, 1968 a)

eyes were shut but the withdrawal reflex to toe pinching was present, dorsiflexion of the neck and extension of the legs were observed following an intravenous injection of PGF$_{2\alpha}$. In two chicks more deeply anaesthetised with urethane (2.25 g/kg) the withdrawal reflex was absent and PGF$_{2\alpha}$ (40 µg/kg) had no effect (HORTON and MAIN, 1967 a).

g) Effect of PGF$_{2\alpha}$ on the Spinal Cord

In chicks anaesthetised with urethane or chloralose PGF$_{2\alpha}$ injected intravenously increases gastrocnemius muscle tension measured isometrically (Fig. 10). The increase in tension is similar in onset and duration to the extension of the legs observed in unanaesthetised chicks. The contraction begins 20 to 30 seconds after the injection and

Fig. 9. Chick 39 g photographed 1 minute after an intravenous injection of
4 µg PGF$_{2\alpha}$ (HORTON and MAIN, 1967 b)

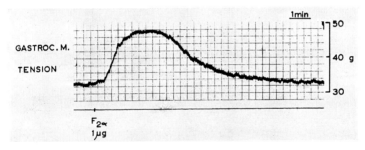

Fig. 10. Chicken 550 g anaesthetised with urethane (1.7 g/kg). Effect of
1 µg of PGF$_{2\alpha}$ injected intravenously on gastrocnemius muscle tension
recorded isometrically. (HORTON and MAIN, 1967 a)

lasts up to 10 minutes. The sensitivity to PGF$_{2\alpha}$ (on a body weight
basis) does not decrease with age and it is similar in chicks anaesthetised with either urethane or chloralose. In 9 chicks the lowest effective dose of PGF$_{2\alpha}$ ranged from 2 to 100 µg/kg (HORTON and MAIN,
1967 a). Tachyphylaxis to successive doses of PGF$_{2\alpha}$ often developes
if the interval between doses is short. Tachyphylaxis to the action of
PGF$_{2\alpha}$ has also been observed on application to single neurones in the
brain stem (AVANZINO et al., 1966).

Defaecation and opening of the eyes sometimes accompanied the
increase in gastrocnemius muscle tension. Eye opening in these anaesthetised chicks was recorded by measuring the tension of the lower

eyelid isometrically. The time course of increase in eyelid tension in
response to $PGF_{2\alpha}$ approximately paralleled the increase in gastrocnemius muscle tension (Fig. 11).

These experiments confirm that the extension of the legs observed
in unanaesthetised chicks after $PGF_{2\alpha}$ can be attributed to contraction
of extensor muscles such as the gastrocnemius. In one unanaesthetised
chick the left sciatic nerve was sectioned under local anaesthesia.

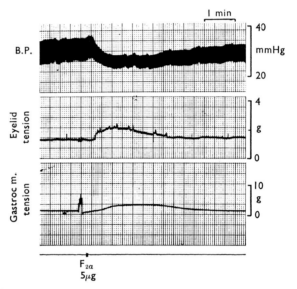

Fig. 11. Chick 85 g anaesthetised with urethane (2 g/kg). Upper trace:
systemic arterial blood pressure. Middle trace: lower eyelid tension. Lower
trace: gastrocnemius muscle tension. Responses to $PGF_{2\alpha}$ (5 µg) injected
intravenously (HORTON and MAIN, 1967 a)

Fifteen minutes after section an intravenous injection of $PGF_{2\alpha}$
(330 µg/kg) resulted in the usual dorsiflexion of the neck accompanied
by extension of the right (innervated) limb only. There was no extension of the left *(denervated)* limb. In the anaesthetised chick $PGF_{2\alpha}$
also failed to produce contraction if the gastrocnemius muscle was
acutely denervated. Since the effect of $PGF_{2\alpha}$ is abolished by de-
nervating the muscle, it must be dependent upon central pathways.
The site of action of $PGF_{2\alpha}$ was located further as follows (HORTON
and MAIN, 1967 a):

In decapitated and artifically ventilated chicks whose spinal cord has been sectioned in the mid-cervical region, PGF$_{2\alpha}$ (9 to 166 µg/kg) injected intravenously, still increases gastrocnemius muscle tension. Clearly the brain cannot be necessary for the mediation of this effect nor can it be a reflex response mediated by pathways which depend upon the integrity of the brain stem.

In the experiment on a spinal chick illustrated in Fig. 12, both sciatic nerves were exposed and the tension of both gastrocnemius

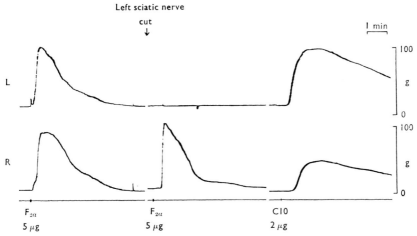

Fig. 12. Spinal chick (decapitated weight 32 g). Gastrocnemius muscle tension recorded isometrically (upper tracing left leg, lower tracing right leg). Responses to intravenous injections of PGF$_{2\alpha}$ (5 µg) and decamethonium iodide (C 10, 2 µg). Between the first and second panel the left sciatic nerve was cut. There was an interval of 60 minutes between each injection. (HORTON and MAIN, 1967 a)

muscles was recorded. PGF$_{2\alpha}$ (156 µg/kg) injected intravenously increased the tension of both the left and the right gastrocnemius muscle. After the *left* sciatic nerve had been sectioned the same dose of PGF$_{2\alpha}$ injected 1 hour later increased the tension of the *right* gastrocnemius muscle only. Later decamethonium iodide (63 µg/kg) was injected intravenously and caused bilateral contraction showing that the denervated gastrocnemius muscle was still capable of contracting in response to drugs injected intravenously. These experiments established that the response of the gastrocnemius muscle to PGF$_{2\alpha}$ is not

extension of the forelimbs and walking movements. There is no further evidence about the site of action of $PGF_{2\alpha}$ in the cat though this compound and $PGF_{1\alpha}$ produce long-lasting changes in monosynaptic reflexes in the chloralose-anaesthetised cat with intact brain (Fig. 15) (DUDA, HORTON and MCPHERSON, 1968).

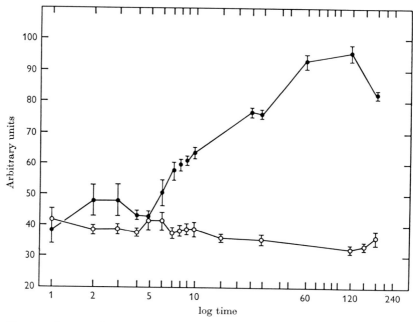

Fig. 15. Cat (2.8 kg) anaesthetised with chloralose. The open circles represent reflex responses measured over a 3 hour period following an intra-aortic injection of 10 µg $PGF_{1\alpha}$. The filled circles represent reflex responses measured later in the same cat after an injection of 55 µg $PGF_{1\alpha}$. Ordinate: amplitude of reflex response in arbitrary units. Abscissa: time in minutes after the injections (log scale). (DUDA, HORTON and MCPHERSON, 1968)

h) Effect of PGE_1 on Motor Pathways in the Brain and Spinal Cord

The effect of PGE_1 on central pathways is more complex. In the unanaesthetised chick, PGE_1 injected intravenously does not cause gastrocnemius muscle contraction but in the lightly urethanised chick PGE_1 causes a *reduction* in gastrocnemius muscle tension (Fig. 16).

In the chloralosed chick with intact brain, PGE_1 (1—8 µg/kg) either has no effect or *inhibits* the crossed extensor reflex. In the spinal chick, however, PGE₁ in the same dose potentiates the crossed extensor reflex (HORTON and MAIN, 1967 a).

This suggests that PGE₁ may have at least two sites of action, one at the spinal level which results in potentiation of the crossed extensor reflex and a second on higher centres in the brain resulting, often, in inhibition of the reflex.

PGE₁ injected intravenously (10—20 μg/kg) in the decerebrate cat, like PGF$_{2\alpha}$, potentiates decerebrate rigidity. It increases gastrocnemius muscle tension (Fig. 14), an effect which is prevented by

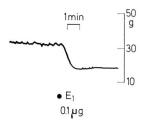

Fig. 16. Chicken 550 g anaesthetised with urethane. Gastrocnemius muscle tension recorded isometrically. PGE₁ (0.1 μg) injected intravenously reduced tension and abolished tremor. (Experiment by HORTON and MAIN)

acute denervation of the muscle. On close arterial injection to the muscle however PGE₁ had no effect, though in higher doses bilateral responses occurred after a longer latent period than following intravenous injection. These results show that the increased gastrocnemius muscle tension following intravenous PGE₁ in the decerebrate cat is *not* due to an action on skeletal muscle or the neuromuscular junction. (This result parallels that found for PGF$_{2\alpha}$ in the chick.)

The increase in gastrocnemius muscle tension in response to PGE₁ was also observed in the spinal cat. The response is abolished by section of the nerve to the muscle but *not* by dorsal root section. Furthermore, as in the decerebrate animal, PGE₁ injected close-arterially had no effect on gastrocnemius muscle tension. It is concluded from these results that the increase in gastrocnemius muscle tension in the decerebrate cat following intravenous PGE₁ must be centrally mediated. Reflex contractions of the muscle mediated *via* the brain could not of course account for the positive responses obtained in the spinal animal but reflex contractions to PGE₁ acting upon spinal afferent nerves are not excluded. A direct action of PGE₁ on the cord is more likely, since contractions of the gastrocnemius muscle have been elicited by topical application of PGE_1 (500 µg/ml) to the lumbar spinal cord. Responses to prostaglandins applied topically to the spinal cord have been recorded in the toad also (PHILLIS and TEBECIS, 1968). Since the effect of PGE_1 on gastrocnemius muscle tension persists after dorsal root section, it cannot be due to facilitation of γ-motoneurone firing alone. It seems probable that PGE_1

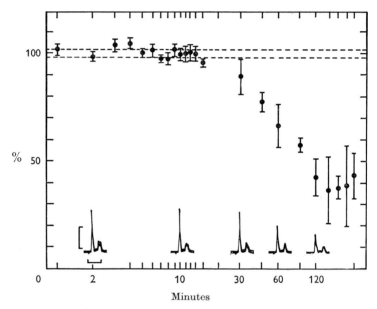

Fig. 17. Cat (2.8 kg) anaesthetised with chloralose. The effect of PGE_1 (10 µg) injected into the aorta on monosynaptic reflex responses. Samples of the action potentials at 2, 10, 30, 60 and 120 minutes after the injection are illustrated. Ordinate: % change in amplitude of monosynaptic reflex response. Abscissa: time in minutes (log scale) after the injection. Calibration for the action potential: 0.5 mV, 10 msec. (DUDA et al., 1968)

facilitates the firing of α-motoneurones either directly or indirectly, possibly by stimulation of excitatory pathways since PGE_1 did not affect inhibition of the crossed extensor reflex produced by electrical stimulation of the ipsilateral peroneal nerve (HORTON and MAIN, 1967 a). These conclusions do not exclude the likely possibility that PGE_1 has, in addition, actions upon brain stem descending facilitatory pathways, which contribute to the responses observed in the decerebrate animal.

Similar conclusions were reached by HORTON and MAIN (1969) as a result of their work on cats. The patellar reflex in chloralosed cats is usually inhibited (in one experiment facilitation occurred) confirming the results obtained by recording evoked potentials in the ventral roots (Fig. 17) (DUDA et al., 1968). This inhibitory response is of long duration. Crossed extensor reflex responses are also diminished by PGE₁ in the chloralosed cat, although twitches of the tibialis anterior muscle in response to electrical stimulation of its motor nerve are undiminished. The inhibition of the crossed extensor reflex has also been observed in the decerebrate cat, whether anaesthetised with chloralose or not. On the other hand in the spinal animal facilitation of the crossed extensor reflex is the most significant effect (Fig. 18) (HORTON and MAIN, 1967 a) though this may subsequently be overshadowed by inhibition (HORTON and MAIN, 1969).

Thus in the spinal cat as in the spinal chick, PGE₁ appears to have a facilitatory action on spinal reflexes. When higher centres are present, this facilitation is not observed presumably because PGE₁ acts upon supraspinal centres with descending pathways which impinge upon neurones of the spinal reflex pathways.

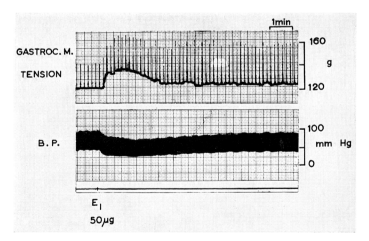

Fig. 18. Spinal cat (2.4 kg). Upper trace: gastrocnemius muscle tension, muscle twitches were elicited by electrical stimulation of the central stump of the contralateral sciatic nerve. Lower trace: carotid arterial blood pressure. (HORTON and MAIN, 1967 a)

It is also apparent from the biphasic responses to PGE₁ obtained in spinal cats (HORTON and MAIN, 1969) as well as from the variable responses in the chloralosed cat with intact brain that PGE₁ almost certainly has numerous sites of action on motor pathways in the central nervous system.

i) Abolition of Tremor by PGE₁

Although PGE₁ does not prevent oxotremorine-induced tremor in mice (HOLMES and HORTON, unpublished), it does abolish pentobarbitone tremor (or shivering) (HORTON and MAIN, 1967 b). It is known that cats recovering from an anaesthetic dose of pentobarbitone frequently exhibit signs of tremor or shivering and that this can be abolished by administration of certain drugs into the cerebral ventricles. HORTON and MAIN (1967 b) had observed that chicks recovering from urethane anaesthesia also exhibited tremor. In the cat and chick, the tremor could be abolished very rapidly by the intravenous injection of PGE₁ (Fig. 19). Tremor in the chick could also be abolished by direct application of PGE₁ to the cerebrum but in the cat neither injection into the lateral cerebral ventricles nor topical application to the cerebral cortex abolished the tremor.

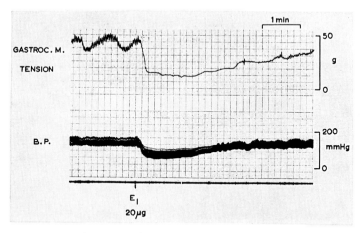

Fig. 19. Cat (3.2 kg) anaesthetised with sodium pentobarbitone. Upper trace: gastrocnemius muscle tension. Lower trace: carotid arterial blood pressure. Responses to PGE₁ (20 µg) injected intravenously. (HORTON and MAIN, 1967 b)

MAIN and WRIGHT (1969) carried out experiments in an attempt to elucidate the site of this anti-tremor action of PGE_1. Injection into the artery supplying the gastrocnemius muscle caused only a delayed inhibition of tremor similar in magnitude and duration to that seen after an intravenous injection. This result shows that PGE_1 is not acting on the muscle or neuro-muscular junction. Neither perfusion of PGE_1 (20 µg/ml) from the lateral ventricles to the cisterna magna nor the topical application of PGE_1 (10—100 µg/ml) to the cerebral cortex, nor injection into the carotid or vertebral arteries or the abdominal aorta in cats provided any evidence for a direct central action. There is some evidence to suggest that the inhibition of tremor may be secondary to changes in cutaneous blood flow induced by PGE_1 (MAIN and WRIGHT, 1969). There is also evidence that PGE_1 can affect the thermoregulatory centre in the hypothalamus (MILTON and WENDLANDT, 1970); this could contribute to the abolition of shivering.

j) Action of Prostaglandins on the Brain Stem

In dogs exhibiting marked sinus arrhythmia, PGE_1 (1.5—10 µg/kg) injected into a common carotid artery abolishes or markedly reduces the arrhythmia, the duration of the cardiac cycle becoming equal to that in the inspiratory phase of the arrhythmia (Fig. 20) (McQUEEN and UNGAR, 1969). Thus PGE_1 appears to inhibit expiratory slowing rather than to produce inspiratory speeding. The effect was not altered by denervation of the carotid sinus region and became apparent within 2.5 seconds of the injection. It is therefore a central action. The arrhythmia returned after 5 to 20 minutes. Tachyphylaxis occurred after 2 or 3 injections repeated at 20 minute intervals.

During the height of the response to PGE_1 stimulation of the carotid body chemoreceptors gave rise to a bradycardia similar to one evoked before the PGE_1 was injected. The pathways for reflex cardiac inhibition are thus not blocked by PGE_1.

In some experiments PGE_1 abolished the central component of the Traube-Hering waves in arterial blood pressure. This effect was transient. There was also an increase in pulmonary ventilation, sometimes as great as 100% due to a reduction in the duration of expiration. This effect was slightly later in onset than the abolition of arrhythmia and it was accompanied by no changes in other respiratory

parameters including duration and amplitude of inspiration and inspiratory and expiratory peak flow. Similar responses, isolated to a
single respiratory parameter, are unusual but have been described on
localised electrical stimulation of the brain stem in cats (HUGELIN
and COHEN, 1963).

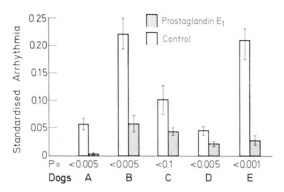

Fig. 20. Inhibition of sinus arrhythmia by PGE$_1$ (1.5—10 µg/kg) in the dog.
Ordinate: standardised arrhythmia = standard deviation of the cardiac
period (= duration of the cardiac cycle, beat by beat) divided by the mean
cardiac period. Open columns: control values. Black columns: values after
PGE$_1$. Abscissa: dogs A to E. (McQEEN and UNGAR, 1969)

It is concluded that PGE$_1$ has a selective action on a group of
neurones within the brain stem, the effect of which is to inhibit the
irradiation of activity from respiratory to cardio-inhibitor and vaso-
motor centres.

Cross-circulation experiments in dogs show that PGE$_1$ (10 µg/kg)
injected into the carotid artery has a central pressor effect which can
be abolished by hexamethonium (KAPLAN, GREGA, SHERMAN and
BUCKLEY, 1969). LAVERY, LOWE and SCROOP (1970) have also described cardiovascular actions of prostaglandins which can be attributed to a site of action in the central nervous system. PGE$_1$ infused
into a vertebral artery at 4 to 360 ng/kg/min caused a tachycardia
which was greater than that produced by intracarotid or intravenous
infusions. PGF$_{2\alpha}$ (4—64 ng/kg/min) infused into the vertebral artery
caused an increase in systemic arterial blood pressure, a fall in central
venous pressure and tachycardia (Fig. 21). Cardiac output was increased but peripheral resistance was unchanged. Since PGF$_{2\alpha}$ in this

dosage was inactive by intravenous infusion it is concluded that the changes are due to an effect on structures within the territory of distribution of the vertebral artery, presumably the central nervous system. Changes in vertebral blood flow cannot account for these responses (LAVERY et al., 1970).

When $PGF_{2\alpha}$ (10 µg/ml) is applied by local superfusion to the surface of the medulla, synaptic transmission in the cuneate nucleus

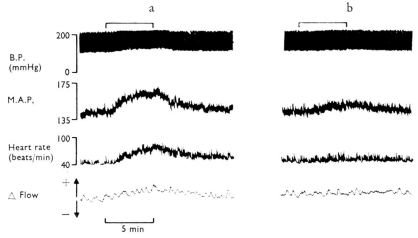

Fig. 21. Effect in a chloralosed dog of $PGF_{2\alpha}$ infusions (a) into the vertebral artery at 0.4 µg/min and (b) intravenously at 50 µg/min. Pulsatile arterial pressure (B. P.), mean arterial pressure (M. A. P.), heart rate and vertebral artery blood flow are illustrated. (LAVERY, LOWE and SCROOP, 1970)

is reduced (COCEANI, DREIFUSS, PUGLISI and WOLFE, 1969). Direct electrical stimulation in the vicinity of the cuneate neurones elicits evoked potentials which are conducted orthodromically within the median lemniscus and antidromically along peripheral nerves. $PGF_{2\alpha}$ reduced all components of the orthodromic response whilst the antidromic response was unaffected (Fig. 22). PGE_1 was inactive on this system in the dose tested (COCEANI, 1971).

k) Micro-iontophoretic Studies

The work of AVANZINO et al. (1966) established that prostaglandins have a direct action on central neurones. The results of their studies on brain stem neurones in the decerebrate cat are summarised

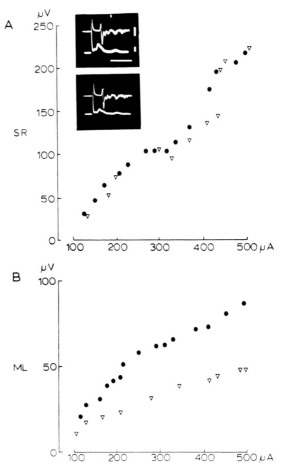

Fig. 22. The effect of PGF$_{2\alpha}$ (10 µg/ml) superfused over the medulla oblongata of a cat, on antidromic (insert upper trace) and orthodromic (insert lower trace) responses evoked by direct electrical stimulation of the cuneate nucleus. Antidromic response recorded in the superficial radial nerve (S. R.), orthodromic response in the medial lemniscus (M. L.). Upper insert, control; lower insert, PGF$_{2\alpha}$. Calibration: 100 µV, 5 msec. A. shows relation between intensity of stimulation and amplitude of the antidromic response. B. shows same as A. for orthodromic response. ● control; ▽ PGF$_{2\alpha}$. (Coceani et al., 1969)

in Table 1. Tachyphylaxis or desensitisation was frequently observed but there was, significantly, no cross tachyphylaxis between different prostaglandins (Fig. 23).

Table 1. *Action of prostaglandins on spontaneously firing brain stem neurones in the decerebrate cat* (Avanzino et al., 1966)

Response	PGE_1	PGE_2	$PGF_{2\alpha}$
Excitatory	89	19	40
Inhibitory	9	0	15
No effect	243	50	100
Total neurones tested	341	69	155

More recently Hoffer, Siggins and Bloom (1969) have demonstrated that PGE_1 and PGE_2 antagonise the inhibitory effects of noradrenaline on the spontaneous firing of Purkinje cells in the cerebellum. Furthermore, though cyclic AMP mimics the action of noradrenaline, the effect of cyclic AMP is *not* reversed by the prostaglandins (Fig. 24). It is postulated that PGE compounds produced locally in the Purkinje cells antagonise the action of noradrenaline by blocking its effects on adenyl cyclase (see Chapter XV) (Siggins, Hoffer and Bloom, 1969). The histochemical localisation of prostaglandin dehydrogenase in the Purkinje cell layer of the cerebellum (Siggins, Hoffer and Bloom, 1971) provides additional evidence that prostaglandins have a physiological role in this region.

l) Conclusions

It is apparent that the widespread distribution of prostaglandins in the central nervous system (Chapter II) is paralleled by numerous sites of actions of these compounds in the central nervous system.

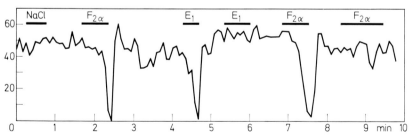

Fig. 23. Decerebrate cat. Effects of PGE_1 (100 nA) and $PGF_{2\alpha}$ (100 nA) on the impulse frequency of a neurone in the nucleus reticularis gigantocellularis. Desensitisation occurred with repeated doses. (Avanzino, Bradley and Wolstencroft, 1966)

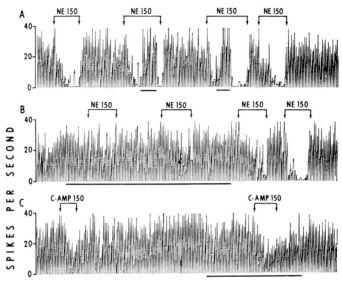

Fig. 24. Selective antagonism by PGE_1 of Purkinje cell responses to noradrenaline (cerebellum of the rat under chloral hydrate). A, B and C
represent consecutive records from the same cell. Duration of noradrenaline
(NE) and cyclic AMP micro-electrophoresis indicated by arrows. Numbers
after each drug indicate ejection current in nanoamps. Black lines indicate
PGE_1, 125 nA. (HOFFER, SIGGINS and BLOOM, 1969)

Although the role of prostaglandins in the brain and spinal cord is
not yet clear, certain general observations can be made: (1) All prostaglandins do not affect all neurones or all pathways. (2) Tachyphylaxis to central actions is a common finding though there appears
to be no cross tachyphylaxis. (3) The duration of action of the prostaglandins is often long and may exceed the time the compounds are
actually in contact with their site of action. (4) Central effects can
sometimes be produced by very small doses of prostaglandins in the
circulation, yet it is known from radioactive studies that only a
minute proportion will actually reach the brain or spinal cord. No
radioactivity can be detected in the brain by autoradiographic techniques following intravenous or subcutaneous administration of PGE_1
or $PGF_{2\alpha}$ (see Chapter V). Moreover even when tritiated PGE_1 is injected into the carotid or vertebral arteries or into the ventricles, very
little radioactivity can be detected in the brain removed at a time
when the central actions of the PGE_1 are still maximal (HOLMES and
HORTON, 1968 b).

2. Peripheral Nervous System

a) Sensory Endings

Antidromic stimulation of the trigeminal nerve causes a long-lasting miosis and vasodilatation, both of which still occur after pretreatment with atropine. Since prostaglandins occur in the iris (see Chapter II) it is of interest that PGE_1 and PGE_2 (0.01—0.1 µg) cause miosis and an elevation of intra-ocular pressure when injected into the anterior chamber of the rabbit eye (WAITZMAN and KING, 1967; BEITCH and EAKINS, 1969). Atropine-resistant miosis is also observed in the cat after intra-ocular injection of PGE_1 or PGE_2. It seems possible that the release of prostaglandins may account for the effects observed on antidromic stimulation of trigeminal nerve.

Sensory nerve endings in human skin do not respond to PGE_1 except in high concentration (100 µg/ml), in contrast to other smooth muscle stimulants such as 5-hydroxytryptamine and bradykinin which cause a sensation of pain when applied to an exposed blister base on the human arm (HORTON, 1963). The prostaglandin precursors, di-homo-γ-linolenic and arachidonic acid, are slightly more active than PGE_1 in this test (HORTON and MAIN, 1966).

There is evidence that the depressor action of PGE_1 can be attributed at least in part to an action on receptors in the region of the carotid bifurcation (KAPLAN et al., 1969). The depressor response to an injection of PGE_1 into the common carotid artery can be abolished or reversed by cutting the sinus nerve.

b) Autonomic Ganglia and Neuro-muscular Junction

PGE_1 had no detectable action on synaptic transmission through the superior cervical ganglion (KAYAALP and McISAAC, 1968). In high doses only (20—30 µg/kg) PGE_1 injected close-arterially to a gastrocnemius muscle in the cat reduces the sizes of twitches evoked by stimulating the motor nerve. In one experiment on the chronically denervated gastrocnemius muscle, HORTON and MAIN (1967 a) found that contractures in response to intra-arterial acetylcholine were reduced in size after a single injection of PGE_1 though the effects of direct electrical stimulation of the muscle were unchanged.

c) Adrenergic Neurones

DAVIES, HORTON and WITHRINGTON (1968) identified PGE_2 in venous effluent from the blood-perfused spleen following splenic nerve stimulation. However this compound when injected intra-arterially has little pharmacological activity on either capsular or vascular smooth muscle of the dog spleen or their response to catecholamines (DAVIES and WITHRINGTON, 1968).

During $PGF_{2\alpha}$ infusion (2 μg/min) which is also released from the spleen (GILMORE, VANE and WYLLIE, 1968) there was a small but consistent increase in the size of splenic contraction produced by nerve stimulation. In contrast, the contractions due to noradrenaline were unaffected by $PGF_{2\alpha}$ (DAVIES and WITHRINGTON, 1969).

In the cat spleen Hedqvist has elegantly demonstrated that PGE_2 reduces the response to splenic nerve stimulation and reduces the output of noradrenaline (HEDQVIST, 1969 a, 1969 b, 1969 c, 1970). It may thus be postulated that PGE_2 released from the post-synaptic site in response to nerve stimulation acts at a pre-synaptic site to reduce the amount of transmitter released in response to continued nerve stimulation. The possibility that this mechanism might be a more generalised one applying also to sites in the central nervous system must be investigated.

References

AVANZINO, G. L., BRADLEY, P. B., WOLSTENCROFT, J. H.: Actions of pro-staglandins E_1, E_2 and $F_{2\alpha}$ on brain stem neurones. Br. J. Pharmac. **27**, 157—163 (1966).

BEITCH, B. R., EAKINS, K. E.: The effects of prostaglandins on the intra-ocular pressure of the rabbit. Br. J. Pharmac. **37**, 158—167 (1969).

BIANCHI, C., FRANCESCHINI, J.: Experimental observations on Haffner's method for testing analgesic drugs. Br. J. Pharmac. **9**, 280—284 (1954).

COCEANI, F., DREIFUSS, J. J., PUGLISI, L., WOLFE, L. S.: Prostaglandins and membrane function. In: Prostaglandins, Peptides and Amines. Ed.: P. MANTEGAZZA and E. W. HORTON. London: Academic Press 1969, pp. 73—84.

— PUGLISI, L., LAVERS, B. H.: Prostaglandins and synaptic activity in spinal cord and cuneate nucleus. Ann. N. Y. Acad. Sci. **80**, 289—300 (1971).

COLLIER, H. O. J.: Analgesics. Evaluation of drug activities: Pharmacometrics, Vol. 1. Ed.: D. R. LAURENCE and A. L. BACHARACH. London: Academic Press 1964, pp. 183—204.

— HAMMOND, A. R., HORWOOD-BARRETT, S., SCHNEIDER, C.: Rapid induction by acetylcholine, bradykinin and potassium of a nociceptive response in mice and its selective antagonism by aspirin. Nature 204, 1316—1318 (1964).

DAVIES, B. N., HORTON, E. W., WITHRINGTON, P. G.: The occurrence of prostaglandin E_2 in splenic venous blood of the dog following splenic nerve stimulation. Br. J. Pharmac. 32, 127—135 (1968).

— WITHRINGTON, P. G.: The effects of prostaglandin E_1 and E_2 on the smooth muscle of the dog spleen and on its responses to catecholamines, angiotensin and nerve stimulation. Br. J. Pharmac. 32, 136—144 (1968).

— — Actions of prostaglandins A_1, A_2, E_1, E_2 and $F_1\alpha$ and $F_2\alpha$ on splenic vascular and capsular smooth muscle and their interactions with sympathetic nerve stimulation, catecholamines and angiotensin. In: Prostaglandins, Peptides and Amines. Ed.: P. MANTEGAZZA and E. W. HORTON. London: Academic Press 1969, pp. 53—56.

DUDA, P., HORTON, E. W., McPHERSON, A.: The effects of prostaglandins E_1, $F_{1\alpha}$ and $F_{2\alpha}$ on monosynaptic reflexes. J. Physiol. (Lond.) 196, 151—162 (1968).

DURU, S., TÜRKER, R. K.: Effect of prostaglandin E_1 on the strychnine-induced convulsion in the mouse. Experientia 25, 275 (1969).

FELDBERG, W., SHERWOOD, S. L.: Behaviour of cats after intraventricular injections of eserine and DFP. J. Physiol. (Lond.) 125, 488—500 (1954).

— — Injections of bulbocapnine into the cerebral ventricles of cats. Br. J. Pharmac. 10, 371—374 (1955).

GILMORE, N., VANE, J. R., WYLLIE, J. H.: Prostaglandin released by the spleen. Nature, 218, 1135—1140 (1968).

HEDQVIST, P.: Modulating effect of prostaglandin E_2 on noradrenaline release from the isolated cat spleen. Acta physiol. scand. 75, 511—512 (1969 a).

— Antagonism between prostaglandin E_2 and phenoxybenzamine on noradrenaline release from the cat spleen. Acta physiol. scand. 76, 383—384 (1969 b).

— Prostaglandin E as a modulator of noradrenaline release from sympathetic nerves. Acta physiol. scand. 77, suppl. 330, 59 (1969 c).

— Control by prostaglandin E_2 of sympathetic neurotransmission in the spleen. Life Sci. 9, part 1, 269—278 (1970).

HOFFER, B. J., SIGGINS, G. R., BLOOM, F. E.: Prostaglandins E_1 and E_2 antagonize norepinephrine effects on cerebellar Purkinje cells: Microelectrophoretic study. Science, N. Y. 166, 1418—1420 (1969).

HOLMES, S. W.: Prostaglandins in the central nervous system. Ph.D. Thesis. University of London 1968.

— Horton, E. W.: Prostaglandins and the central nervous system. Prostaglandin Symposium of the Worcester Foundation for Exp. Biol. Ed.: P. W. Ramwell and J. E. Shaw. New York: Interscience 1968 a, pp. 21—38.

— — The distribution of tritium-labelled prostaglandin E_1 injected in amounts sufficient to produce central nervous effects in cats and chicks. Br. J. Pharmac. 34, 32—37 (1968 b).

Horton, E. W.: Action of prostaglandin E_1 on tissues which respond to bradykinin. Nature 200, 892—893 (1963).

— Actions of prostaglandins E_1, E_2 and E_3 on the central nervous system. Br. J. Pharmac. 22, 189—192 (1964).

— Main, I. H. M.: Differences in the effects of prostaglandin $F_{2\alpha}$, a constituent of cerebral tissue, and prostaglandin E_1 on conscious cats and chicks. Int. J. Neuropharmac. 4, 65—69 (1965) [see erratum 4, 359 (1965)].

— — The relationship between the chemical structure of prostaglandins and their biological activity. Mem. Soc. Endocr. 14, 29—36 (1966).

— — Further observations on the central nervous actions of prostaglandins $F_{2\alpha}$ and E_1. Br. J. Pharmac. 30, 568—581 (1967 a).

— — Central nervous actions of the prostaglandins and their identification in the brain and spinal cord. Nobel Symposium 2, Prostaglandins. Ed.: S. Bergström and B. Samuelsson. Stockholm: Almqvist and Wiksell 1967 b, pp. 253—260.

— — Actions of prostaglandin E_1 on spinal reflexes in the cat. In: Prostaglandins, Peptides and Amines. Ed.: P. Mantegazza and E. W. Horton. London: Academic Press 1969, pp. 121—122.

Hugelin, A., Cohen, M. I.: The reticular activating system and respiratory regulation in the cat. Ann. N. Y. Acad. Sci. 109, 586—603 (1963).

Kaplan, H. R., Grega, G. J., Sherman, G. P., Buckley, J. P.: Central and reflexogenic cardiovascular actions of prostaglandin E_1. Int. J. Neuropharmac. 8, 15—24 (1969).

Kayaalp, S. O., McIsaac, R. J.: Absence of effects of prostaglandins E_1 and E_2 on ganglionic transmission. Eur. J. Pharmac. 4, 283—288 (1968).

Lavery, H. A., Lowe, R. D., Scroop, G. C.: Cardiovascular effects of prostaglandins mediated by the central nervous system of the dog. Br. J. Pharmac. 39, 511—519 (1970).

Main, I. H. M., Wright, P. M.: The abolition of tremor by prostaglandin E_1. In: Prostaglandins, Peptides and Amines. Ed.: P. Mantegazza and E. W. Horton. London: Academic Press 1969, pp. 125—127.

McQueen, D. S., Ungar, A.: The modification by prostaglandin E_1 of central nervous interaction between respiratory and cardioinhibitory pathways. In: Prostaglandins, Peptides and Amines. Ed.: P. Mantegazza and E. W. Horton. London: Academic Press 1969, pp. 123—124.

Milton, A. S., Wendlandt, S.: A possible role for prostaglandin E_1 as a modulator for temperature regulation in the central nervous system of the cat. J. Physiol., (Lond.) 207, 76P—77P (1970).

PHILLIS, J. W., TEBĒCIS, A. K.: Prostaglandins and toad spinal cord responses. Nature, **217**, 1076—1077 (1968).

SIGGINS, G. R., HOFFER, B. J., BLOOM, F. E.: Cyclic adenosine monophosphate: Possible mediator for norepinephrine effects on cerebellar Purkinje cells. Science, N. Y. **165**, 1018—1020 (1969).

— — — Specificity of prostaglandin E_1-norepinephrine antagonisms in the brain: microelectrophoretic and histochemical correlations. Ann. N. Y. Acad. Sci. **80**, 302—319 (1971).

WAITZMAN, M. B., KING, C. D.: Prostaglandin influences on intraocular pressure and pupil size. Am. J. Physiol. **212**, 329—334 (1967).

X. Cardiovascular System

Prostaglandins act upon systemic blood vessels, the pulmonary vascular bed, the heart, carotid sinus and central nervous pathways controlling the cardiovascular system; moreover their effects differ according to which prostaglandin and which species are being studied. No attempt will be made in this chapter to cover the vast literature which has been devoted to this field.

PGE_1 and PGE_2 lower systemic arterial blood pressure on intravenous injection in all species so far investigated. PGE_1 increases blood flow through various vascular beds on intra-arterial injection; for example in the anaesthetised dog PGE_1 (0.1 µg/kg) injected into the brachial, femoral, carotid, coronary and renal arteries decreases peripheral resistance without significant change in systemic arterial blood pressure (NAKANO and McCURDY, 1967; NAKANO and COLE, 1969). Similarly, HORTON and MAIN (1963) observed an increase in blood flow through the hind limb muscle and skin of the anaesthetised cat on intra-arterial injection of PGE_1 and PGE_2, and PGE_1 in doses of 0.1 ng injected close-arterially increases blood flow through canine adipose tissue (FREDHOLM, ÖBERG and ROSELL, 1967). The vasodilator response, like the depressor effect, is unaffected by atropine, propranolol or antihistamine drugs (SMITH, McMORROW, COVINO and LEE, 1968). In contrast to this vasodilator action, PGE_1 constricts nasal mucosal blood vessels in the dog (JACKSON and STOVALL, 1968).

MAXWELL (1967) has also observed in the dog a decrease in systemic arterial pressure during intravenous infusions of PGE_1, accompanied by decreased systemic and pulmonary vascular resistance. PGE_1 had a biphasic effect on system venous return and portal venous pressure, an initial increase being followed by a decrease.

The depressor action of PGA_1 and PGA_2 can probably be attributed to their peripheral vasodilator actions (HORTON and JONES, 1969). In some species (dog, rat and spinal chick) $PGF_{2\alpha}$ increases arterial blood pressure. Since $PGF_{2\alpha}$ is a powerful venoconstrictor, increase in venous return and consequently cardiac output may account for its pressor action (DuCHARME and WEEKS, 1967; DuCHARME, WEEKS and MONTGOMERY, 1968).

Another effect of $PGF_{2\alpha}$ on veins is an antagonism of increased venular permeability to plasma protein induced by histamine, 5-hydroxytryptamine and bradykinin (WILLOUGHBY, 1968). PGE_1 and PGE_2 increase capillary permeability in guinea-pigs (HORTON 1963) and in rats, an effect which, in the rat, is reduced by mepyramine and is probably mediated in part by the release of histamine (KALEY and WEINER, 1968; CRUNKHORN and WILLIS, 1969).

8-iso-PGE_1 which can be formed enzymatically from di-homo-γ-linolenic acid (Chapter IV) is biologically less active than PGE_1 except on pulmonary vascular smooth muscle; its pulmonary vasoconstrictor potency is about 5 times greater than PGE_1 (NAKANO and KESSINGER, 1970).

Contractility of the isolated heart has been studied in various species. PGE_1 increases contractility in the guinea-pig and frog (BERTI, LENTATI and USARDI, 1965); $PGF_{1\alpha}$ has a similar action in the rat (VERGROESEN, DE BOER and GOTTENBOS, 1967). These effects are in general potentiated by lowering calcium concentrations and increasing potassium concentrations in the perfusion fluid (MANTEGAZZA, 1965; VERGROESEN and DE BOER, 1968; PICCININI, POMARELLI and CHIARRA, 1969). The action of PGE_1 on the guinea pig heart appears to be associated with an increase in membrane permeability to calcium (KLAUS and PICCININI, 1967).

PGE_1 (0.2—0.7 µg/kg/min) infused for 4—10 minutes in 2 healthy male subjects caused tachycardia, facial flushing, headache and an oppressive sensation in the chest. There was a moderate fall in cardiac output and systemic arterial blood pressure (BERGSTRÖM, DUNÉR, VON EULER, PERNOW and SJÖVALL, 1959). Furthermore the bradycardia induced by noradrenaline in the human subject is completely inhibited by PGE_1 infusion (BERGSTRÖM, CARLSON, EKELUND and ORÖ, 1965).

$PGF_{2\alpha}$ (0.01—2.0 µg/kg/min) had no effect on systolic or diastolic blood pressure, on heart rate, electrocardiogram or respiratory rate in six subjects (5 male and 1 female). Moreover single rapid injections of up to 40 µg had no effect in these volunteers (KARIM, SOMERS and HILLIER, 1969). Cardiovascular side effects have not been prominent during the infusion of either PGE_2 or $PGF_{2\alpha}$ for the termination of pregnancy (see Chapter VI) nor following oral administration of PGE$_1$ (HORTON, MAIN, THOMPSON and WRIGHT, 1968).

Intradermal injections of PGE$_1$ and PGE$_2$ (50—100 ng) in man induce local oedema and redness lasting 1 to 2 hours (CRUNKHORN and WILLIS, 1969). The possible implication of prostaglandins in cutaneous and other inflammatory disorders must be considered.

Clinically considerable attention has been devoted to the antihypertensive action of PGA compounds and to the possible role of prostaglandins in the control of normal blood pressure. The use of PGF$_{2\alpha}$ as a venoconstrictor in the treatment of shock should also be investigated.

References

BERGSTRÖM, S., CARLSON, L. A., EKELUND, L. G., ORÖ, L.: Cardiovascular and metabolic response to infusions of prostaglandin E$_1$ and to simultaneous infusions of noradrenaline and prostaglandin E$_1$ in man. Acta physiol. scand. **64**, 332—339 (1965).
— DUNÉR, H., VON EULER, U. S., PERNOW, B., SJÖVALL, J.: Observations on the effects of infusion of prostaglandin E in man. Acta physiol. scand. **45**, 145—151 (1959).
BERTI, F., LENTATI, R., USARDI, M. M.: The species specificity of prostaglandin E$_1$ effects on iolated heart. Med. Pharmac. exp. **13**, 233—240 (1965).
CRUNKHORN, P., WILLIS, A. L.: Actions and interactions of prostaglandins administered intradermally in rat and in man. Brit. J. Pharmac. **36**, 216P—217P (1969).
DUCHARME, D. W., WEEKS, J. R.: Cardiovascular pharmacology of prostaglandin F$_{2\alpha}$, a unique pressor agent. Nobel Symposium 2, Prostaglandins, Eds.: S. BERGSTRÖM and B. SAMUELSSON. Stockholm: Almqvist and Wiksell 1967, pp. 173—181.
— — MONTGOMERY, R. G.: Studies on the mechanism of the hypertensive effect of prostaglandin F$_{2\alpha}$. J. Pharmac. exp. Ther. **160**, 1—10 (1968).
FREDHOLM, B. B., OBERG, B., ROSELL, S.: Vascular reactions in canine subcutaneous adipose tissue following prostaglandin E$_1$ (PGE$_1$). Acta pharmac. tox. **25**, suppl. 4, 28 (1967).
HORTON, E. W.: Action of prostaglandin E$_1$ on tissues which respond to bradykinin. Nature (Lond.) **200**, 892—893 (1963).
— JONES, R. L.: Prostaglandins A$_1$, A$_2$ and 19-hydroxy A$_1$, their actions on smooth muscle and their inactivation on passage through pulmonary and hepatic portal vascular beds. Brit. J. Pharmac. **37**, 705—722 (1969).
— MAIN, I. H. M.: A comparison of the biological activities of four prostaglandins. Brit. J. Pharmacol. **21**, 182—189 (1963).

— — Thompson, C. J., Wright, P. M.: Effect of orally administered prostaglandin E_1 on gastric secretion and gastrointestinal motility in man. Gut 9, 655—658 (1968).

Jackson, R. T., Stovall, R.: Vasoconstriction of nasal blood vessels induced by prostaglandins. Prostaglandin Symposium of the Worcester Foundation for Exp. Biol., Eds.: P. W. Ramwell and J. E. Shaw. New York: Inter-science 1968, pp. 329—334.

Kaley, G., Weiner, R.: Microcirculatory studies with prostaglandin E_1. Prostaglandin Symposium of the Worcester Foundation for Exp. Biol. Eds.: P. W. Ramwell and J. E. Shaw. New York: Inter-science 1968, pp. 321—328.

Karim, S. M. M., Somers, K., Hillier, K.: Cardiovascular actions of prostaglandin $F_{2\alpha}$ infusion in man. Eur. J. Pharmac. 5, 117—120 (1969).

Klaus, W., Piccinini, F.: Über die Wirkung von Prostaglandin E_1 auf den Ca-Haushalt isolierter Meerschweinchenherzen. Experientia 23, 556—557 (1967).

Mantegazza, P.: La prostaglandina E_1 come sostanza sensibilizzatrice per il calcio a livello del cuore isolato di cavia. Atti Accad. med. lomb. 20, 66—72 (1965).

Maxwell, G. M.: The effect of prostaglandin E_1 upon the general and coronary haemodynamics and metabolism of the intact dog. Brit. J. Pharmac. 31, 162—168 (1967).

Nakano, J., Cole, B.: Effects of prostaglandins E_1 and $F_{2\alpha}$ on systemic, pulmonary, and splanchnic circulations in dogs. Amer. J. Physiol. 217, 222—227 (1969).

— Kessinger, J. M.: Effects of 8-isoprostaglandin E_1 on the systemic and pulmonary circulations in dogs. Proc. Soc. exp. Biol. Med. 133, 1314—1317 (1970).

— McCurdy, J. R.: Cardiovascular effects of prostaglandin E_1. J. Pharmac. exp. Ther. 156, 538—547 (1967).

Piccinini, F., Pomarelli, P., Chiarra, A.: Further investigations on the mechanisms of the inotropic action of prostaglandin E_1 in relation to the ion balance in frog heart. Pharmac. Res. Commun. 1, 381—389 (1969).

Smith, E. R., McMorrow, J. V., Jr., Covino, B. G., Lee, J. B.: Studies on the vasodilator action of prostaglandin E_1. Prostaglandin Symposium of the Worcester Foundation for Exp. Biol. Eds.: P. W. Ramwell and J. E. Shaw. New York: Inter-science 1968, pp. 259—266.

Vergroesen, A. J., De Boer, J.: Effects of prostaglandins E_1 and $F_{1\alpha}$ on isolated frog and rat hearts in relation to the potassium-calcium ratio of the perfusion fluid. Eur. J. Pharmac. 3, 171—176 (1968).

— — Gottenbos, J. J.: Effects of prostaglandins on perfused isolated rat hearts. Nobel Symposium 2, Prostaglandins. Eds.: S. Bergström and B. Samuelsson. Stockholm: Almqvist and Wiksell 1967, pp. 211—218.

Willoughby, D. A.: Effects of prostaglandins $PGF_{2\alpha}$ and PGE_1 on vascular permeability. J. Path. Bact. 96, 381—387 (1968).

XI. Urinary System

1. Water and Electrolyte Excretion

The increase in water permeability of the toad bladder produced by vasopressin is inhibited by PGE_1 (ORLOFF, HANDLER and BERGSTRÖM, 1965). Similar anti-vasopressin actions of PGE_1 have been observed on the rabbit collecting tubule (ORLOFF and GRANTHAM, 1967). Since these effects of vasopressin are known to be mediated by cyclic AMP, which is not inhibited by PGE_1, it is likely that PGE_1 acts by inhibiting adenyl cyclase.

When PGE_1, PGE_2 or PGA_1 (0.01—2 µg/min) is infused into a renal artery in the hydropenic dog, there is a rise in urinary volume and a rise in urinary sodium, potassium and chloride excretion (HERZOG, JOHNSTON and LAULER, 1968). These doses have no effect upon mean arterial blood pressure or glomerular filtration rate but there is an increase in renal blood flow and a fall in PAH extraction. Even during maximal antidiuresis induced by vasopressin, these prostaglandins cause significant natriuresis and increased free water clearance. It seems probable that the increased urinary volume is mainly secondary to decreased sodium reabsorption rather than to inhibition of vasopressin.

VANDER (1968) substantially confirmed these findings and showed that changes in sodium excretion did not correlate with changes in renal plasma flow or PAH extraction. Using rabbit renal cortical slices, LEE and FERGUSON (1969) demonstrated inhibition of PAH uptake in the presence of several different prostaglandins. This effect is similar to that seen with the postulated natriuretic hormone.

2. Natriuretic Hormone

During the last decade a large volume of evidence has accumulated which tends to support the existence of an unidentified natriuretic hormone. This substance is produced in response to increases in extracellular fluid volume. It inhibits sodium re-absorption, an effect which cannot be accounted for in terms of changes in aldosterone or vasopressin secretion nor changes in glomerular filtration rate. Cross perfusion experiments and assay of plasma samples suggest that this hormone circulates in the blood, but the organ of origin is unknown.

In 1968 NISSEN reported that lipid granules in the interstitial cells of the rat renal medulla decrease in number and size after a few days of salt depletion. Within 30 minutes of repletion with saline, the number and size of granules had been restored to normal. These lipids contain arachidonic acid and since both prostaglandin synthetase and prostaglandin dehydrogenase occur in the kidney, these changes may be associated with changes in prostaglandin turnover in response to alterations in extracellular fluid.

PGE_2, PGA_2 and $PGF_{2\alpha}$ have been identified conclusively in renal medullary tissue (LEE, CROWSHAW, TAKMAN, ATTREP and GOUG-OUTAS, 1967). Since PGA_2 has a potent natriuretic action and since it (unlike PGE_2 and $PGF_{2\alpha}$) is not removed from the circulation by the lungs (HORTON and JONES, 1969; McGIFF, TERRAGNO, STRAND, LEE, LONIGRO and NG, 1969), it has been suggested that the natriuretic hormone is PGA_2 (LEE, 1969). If this hypothesis is true, plasma levels of PGA_2 should be raised in response to expansion of extracellular fluid volume, the usual stimulus to natriuretic hormone output. However, PGA_2 is effective at a blood concentration of 0.1 ng/ml and is difficult to extract in such small quantities. More sensitive methods of identification and estimation are therefore needed to test the hypothesis that natriuretic hormone is a prostaglandin.

References

HERZOG, J. P., JOHNSTON, H. H., LAULER, D. P.: Effects of prostaglandins E_1, E_2 and A_1 on renal hemodynamics, sodium and water excretion in the dog. Prostaglandin Symposium of the Worcester Foundation for Exp. Biol. Eds.: P. W. RAMWELL and J. E. SHAW. New York: Interscience 1968, pp. 147—161.
HORTON, E. W., JONES, R. L.: Prostaglandins A_1, A_2 and 19-hydroxy A_1; their actions on smooth muscle and their inactivation on passage through the pulmonary and hepatic portal vascular beds. Brit. J. Pharmac. 37, 705—722 (1969).
LEE, J. B.: Hypertension, natriuresis, and the renal prostaglandins (Editorial). Ann. intern. Med. 70, 1033—1038 (1969).

— CROWSHAW, K., TAKMAN, B. H., ATTREP, K. A., GOUGOUTAS, J. Z.: The identification of prostaglandins E_2, $F_{2\alpha}$ and A_2 from rabbit kidney medulla. Biochem. J. 105, 1251—1260 (1967).

— FERGUSON, J. F.: Prostaglandins and natriuresis: the effect of renal prostaglandins on PAH uptake by kidney cortex. Nature (Lond.) 222, 1185—1186 (1969).

McGIFF, J. C., TERRAGNO, N. A., STRAND, J. C., LEE, J. B., LONIGRO, A. J., NG, K. K. F.: Selective passage of prostaglandins across the lung. Nature (Lond.) 223, 742—745 (1969).

ORLOFF, J., GRANTHAM, J.: The effect of prostaglandin (PGE_1) on the permeability response of rabbit collecting tubules to vasopressin. Nobel Symposium 2, Prostaglandins. Eds.: S. BERGSTRÖM and B. SAMUELSSON. Stockholm: Almqvist and Wiksell 1967, pp. 143—146.

ORLOFF, J., HANDLER, J. S., BERGSTRÖM, S.: Effect of prostaglandin (PGE_1) on the permeability response of the toad bladder to vasopressin, theophylline and adenosine 3',5'-monophosphate. Nature (Lond.) 205, 397—398 (1965).

VANDER, A. J.: Direct effects of prostaglandin on renal function and renin release in anesthetized dog. Amer. J. Physiol. 214, 218—221 (1968).

XII. Blood Platelets

1. Aggregation in vitro

The aggregation of blood platelets *in vitro* can be measured conveniently by turbidometric methods. On the addition of adenosine diphosphate (ADP) to a suspension of platelets there is a fall in optical density due to platelet aggregation. Although PGE_1 alone has no effect upon platelet aggregation it is a powerful inhibitor of the aggregation induced by ADP (Fig. 1) (KLOEZE, 1967). This effect has been observed on platelets from pig, rat, man and rabbit, and with PGE_1 in concentrations down to 10 ng/ml (Fig. 2). On the other hand PGE_2 has the opposite effect on pig and rat platelets — it enhances the aggregation due to ADP. This striking qualitative difference between two closely related prostaglandins is of considerable interest and has stimulated further work on structure-activity relationship in this field. The carboxyl, 11-oxo and 15-hydroxyl substituents are essential for biological activity, but the C-21 analogue of PGE_1, ω-homo-PGE_1, is almost 4 times more active than the natural C-20 prostaglandin whereas the ω-nor-PGE_1 has only half the biological activity of PGE_1 (KLOEZE, 1969).

PGE_1 also inhibits the adhesiveness of platelets to glass (KLOEZE, 1967) and platelet aggregation induced by other agents (5-hydroxy-tryptamine, thrombin, adenosine triphosphate and noradrenaline) (EMMONS, HAMPTON, HARRISON, HONOUR and MITCHELL, 1967). PGE_1 also inhibits changes in platelet electrophoretic mobility induced by ADP.

Cyclic AMP and its dibutyryl analogue inhibit platelet aggregation (MARQUIS, VIGDAHL and TAVORMINA, 1969) as do the methyl-xanthines which increase cyclic AMP levels by blocking the enzyme phosphodiesterase. PGE_1 increases platelet cyclic AMP concentrations (WOLFE and SHULMAN, 1969) and so it has been suggested that the effects of PGE_1 on platelets can be attributed to changes in cyclic AMP (see Chapter XV).

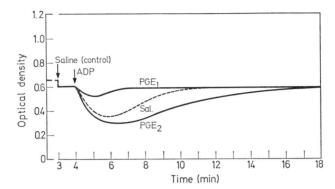

Fig. 1. Platelet aggregation in diluted pig's (citrated) platelet-rich plasma at 37°, after addition of (1), 0.5 μg PGE_1/ml (1.4×10^{-6} M): (2), 0.5 μg PGE_2/ml (1.4×10^{-6} M): (3) saline (control). ADP concentration = 0.6 μg/ml ($\sim 10^{-6}$ M). (KLOEZE, 1967)

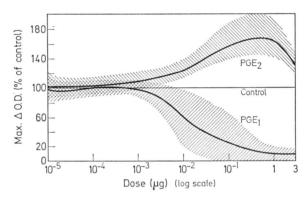

Fig. 2. Dose-response curves, illustrating the effect of PGE_1 and PGE_2 on platelet aggregation induced by ADP at an end concentration of 0.3 μg/ml in rat's (citrated) platelet-rich plasma. The shaded areas indicate the variability. (KLOEZE, 1967)

2. In vivo Experiments

Although PGE_1 is a powerful inhibitor of human platelet aggregation *in vitro*, effects *in vivo* could not be detected when PGE_1 (0.05—0.1 μg/kg/min) was infused intravenously for 30 minutes in 3 healthy male volunteers (CARLSON, IRION and ORÖ, 1968). The concentration of PGE_1 in the circulation was not determined but since prostaglandins are rapidly removed from circulating blood, the plasma levels may well have been lower than in the *in vitro* studies. Higher doses could not be given to man because of side effects.

Platelet aggregation in rats is inhibited by large doses (2 mg/kg) of PGE_1 given intravenously (CHANDRASEKHAR, 1967). Furthermore in rabbits either topical application or intravenous administration (1.6—3.2 µg/kg/min) of PGE_1 reduces platelet thrombus formation in infused blood vessels (EMMONS et al., 1967). Platelet thrombus formation is also significantly reduced in the blood-perfused spleen if PGE_1 is added (BLAKELEY, BROWN, DEARNALEY and WOODS, 1968).

3. Blood Coagulation

Coagulation time of rat blood is not affected by the presence of PGE_1 or PGE_2 whether added to the blood after collection or injected intravenously. Neither prostaglandin has any thromboplastic activity nor do they inhibit thromboplastin. On the other hand PGE_1, added *in vitro*, does reduce the tensile strength of clots, this is presumably attributable to its action on platelets (KLOEZE, 1970).

There was an earlier report that the clotting time of citrated dialysed rat blood is shorter in the presence of PGE_1 for any given calcium ion concentration (FERRI, GALATULAS and PICCININI, 1965). This result could not be confirmed by KLOEZE (1970) using non-dialysed plasma.

References

BLAKELEY, A. G. H., BROWN, G. L., DEARNALEY, D. P., WOODS, R. I.: The use of prostaglandin E_1 in perfusion of the spleen with blood. J. Physiol. (Lond.) **198**, 31—32P (1968).

CARLSON, L. A., IRION, E., ORÖ, L.: Effect of infusion of prostaglandin E_1 on the aggregation of blood platelets in man. Life Sci. **7**, 85—90 (1968).

CHANDRASEKHAR, N.: Inhibition of platelet aggregation by prostaglandins. Blood **30**, 554 (1967).

EMMONS, P. R., HAMPTON, J. R., HARRISON, M. J. G., HONOUR, A. J., MITCHELL, J. R. A.: Effect of prostaglandin E_1 on platelet behaviour *in vitro* and *in vivo*. Brit. med. J. **2**, 468—472 (1967).

FERRI, S., GALATULAS, I., PICCININI, F.: Azione della prostaglandina E_1 sulla coagulazione del sangue. Boll. Soc. ital. Biol. sper. **41**, 1243—1245 (1965).

KLOEZE, J.: Influence of prostaglandins on platelet adhesiveness and platelet aggregation. Nobel Symposium 2, Prostaglandins. Eds.: S. BERGSTRÖM and B. SAMUELSSON. Stockholm: Almqvist and Wiksell 1967, pp 241—252.

KLOEZE, J.: Relationship between chemical structure and platelet-aggregation activity of prostaglandins. Biochim. biophys. Acta 187, 285—292 (1969).

— Influence of prostaglandins E_1 and E_2 on coagulation of rat blood. Experientia 26, 307—308 (1970).

MARQUIS, N. R., VIGDAHL, R. L., TAVORMINA, P. A.: Platelet aggregation. I. Regulation by cyclic AMP and prostaglandin E_1. Biochem. biophys. Res. Commun. 36, 965—972 (1969).

WOLFE, S. M., SHULMAN, N. R.: Adenyl cyclase activity in human platelets. Biochem. biophys. Res. Commun. 35, 265—272 (1969).

XIII. Gastro-intestinal Tract

1. Glandular Secretions in Animals and Man

In the unanaesthetised dog with either an innervated (PAVLOV) or a denervated (HEIDENHAIN) gastric pouch or with a gastric fistula, PGE_1 infused intravenously reduces the gastric secretory response to either food, histamine, pentagastrin or deoxyglucose (Fig. 1). At an infusion rate of 1 µg/kg/min, PGE_1 totally inhibited secretion induced by either food (23 g of canned meat), histamine (1 mg/hr infused intravenously), pentagastrin (75 µg/kg/min infused intravenously), or 2-deoxyglucose (200 mg/kg injected intravenously). At 0.45 µg PGE_1/kg/min there was 60% inhibition. For any dose, maximum inhibition was reached 30 to 45 minutes after the start and was maintained throughout the infusion (ROBERT, NEZAMIS and PHILLIPS, 1967, 1968).

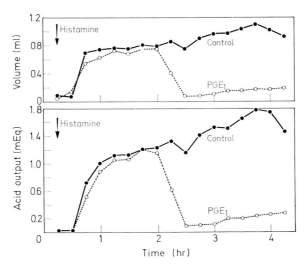

Fig. 1. Effect of PGE_1 (1 µg/kg/min), I.V., on gastric secretion stimulated by histamine infusion (1 mg/hr). Heidenhain pouch. Volume and acid output expressed as per 15 min (ROBERT, NEZAMIS and PHILLIPS, 1967)

PGE$_2$ inhibits the secretory response to food and histamine but has not been tested against pentagastrin and 2-deoxyglucose. PGA$_1$ (0.3 μg/kg/min) inhibits the secretory response to food (Fig. 2).

In the rat anaesthetised with urethane, PGE$_1$ perfused through the lumen of the stomach by the method of GHOSH and SCHILD

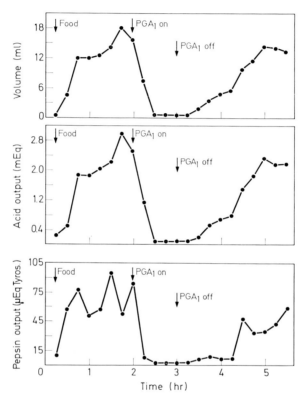

Fig. 2. Effect of PGA$_1$ (0.3 μg/kg/min), I.V., on gastric secretion stimulated with food. Pavlov pouch. Volume and outputs expressed as per 15 min (ROBERT, NEZAMIS and PHILLIPS, 1967)

(1958), reduces the acid output in response to intravenous infusion of histamine or pentagastrin and to vagal stimulation (SHAW and RAMWELL, 1968). PGE$_2$ like PGE$_1$ inhibits acid secretion in the rat and also increases the histamine output from gastric mucosa induced by pentagastrin and histamine (MAIN, 1969). Furthermore in the rat

PGE$_1$ treatment reduces the incidence of ulcer formation and perforation (Fig. 3) (ROBERT, NEZAMIS and PHILLIPS, 1968).

Since PGE$_1$ applied to the mucosal surface is so effective in inhibiting gastric acid secretion in rats, similar inhibition might be expected in man following oral administration. In three healthy male subjects no such inhibition of pentagastrin-induced acid secretion could be detected at least in doses up to 40 μg/kg which produced

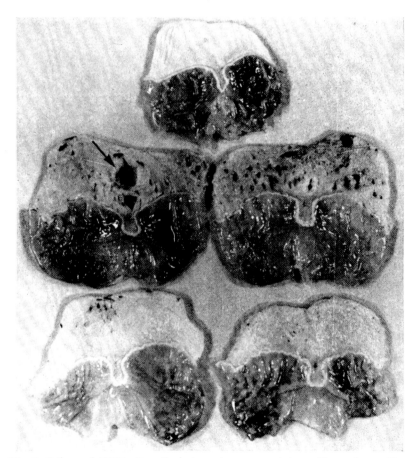

Fig. 3. Effect of PGE$_1$ (2 μg/kg/min, subcutaneously) on Shay ulcers. Top: Intact stomach from nonoperated animal. Middle: Two stomachs with multiple ulcers due to pylorus ligation. One ulcer (arrow) is perforated. Animals were infused with saline. Bottom: Two stomachs with only a trace of ulcerations. Animals were infused with PGE$_1$ and their pylorus was ligated (ROBERT, NEZAMIS and PHILLIPS, 1968)

Table 1. *Effect of prostaglandin E_1 on pentagastrin-induced gastric secretion and on intestinal motility [a] in man*

Exper-iment No.	Subject	Dose (μg E_1/kg)		Vol. (ml)	Acid (m-equiv)	Bile	Increase in Intestinal Motility
		Adminis-tered	Recov-ered				
Control experiments (pentagastrin only)							
1	EH	—	—	184	8.8	0	nil
2	EH	—	—	175	14.5	0	nil
3	HM	—	—	140	11.4	0	nil
4	PW	—	—	148	14.9	0	nil
5 [b]	HM	—	—	N. T.	N. T.	0	nil
Prostaglandin administration							
1	EH	10	3	170	10.0	0	+
2	EH	40	11	241	15.4	+ + +	+ + +
3	HM	20	7	314	24.8	+ + +	+
4	PW	25	13	241	18.0	+ +	+ +
5 [c]	EH	10	0	N. T.	N. T.	N. T.	+ +

[a] The volume (ml) and acidity (m-equiv) refer to the total samples collected during the first hour after pentagastrin injection. N. T. = not tested. Presence of bile in the aspirated samples is indicated by + + or + + +. 0 indicates absence or only trace amounts of bile.

[b] Preliminary experiment to study the side effects of pentagastrin alone.

[c] Preliminary experiment to study the side-effects of prostaglandin E_1 alone (no pentagastrin) (HORTON et al., 1968).

other effects — massive reflux of bile into the stomach and signs of increased intestinal motility (Table 1) (HORTON, MAIN, THOMPSON and WRIGHT, 1968). However, more recently CLASSEN, KOCH, DEYHLE, WEINDENHILLER and DEMLING (1970) have shown that intravenous infusions of PGE_1 (4—5 μg/kg/30 min) in man do inhibit basal gastric acid secretion thus extending the observations on dogs to man. The intravenous route and the incidence of side effects is likely to limit or even prohibit the use of PGE_1 in the treatment of hyperchlorhydria. Nevertheless these early observations hold out hope that a prostaglandin analogue with more specific activity on gastric acid secretion and one which is effective orally may be found.

There is some evidence that the inhibitory action of PGE_1 on gastric secretion may be due to inhibition of adenyl cyclase (SHAW and RAMWELL, 1968; WAY and DURBIN, 1969). However since gastric blood flow is affected by prostaglandins infused intravenously the possibility that the diminution in secretory response is secondary to reduced blood flow must also be considered (WILSON and LEVINE, 1969). JACOBSON (1969) measured the ratio of blood flow to secretion rate in conscious dogs with gastric fistulae and concluded that the inhibition of gastric secretion following PGE_1 is not secondary to vascular changes.

Salivary secretion in the cat in response to chorda tympani stimulation is not inhibited by either PGE_1 or $PGF_{1\alpha}$ (JONES, 1970) injected close arterially. The effect of prostaglandins on salivary blood flow would be of interest since the possible contribution of these substances to the mediation of atropine-resistant salivary vasodilatation must be considered. There have been recent reports that pancreatic secretion is stimulated by prostaglandins (RUDICK, GONDA and JANOWITZ, 1970). Again the mechanism of action may be via adenyl cyclase.

2. Isolated Smooth Muscle Preparations

Isolated segments of gastro-intestinal longitudinal smooth muscle of all species investigated contract in response to prostaglandins E and F (BERGSTRÖM, ELIASSON, VON EULER and SJÖVALL, 1959; HORTON and MAIN, 1963, 1965; BENNETT, ELEY and SCHOLES, 1968 a, 1968 b). Many of these preparations are highly sensitive and have been used extensively since the early work of VON EULER for biological assay of PGE and PGF compounds (see Chapter II).

In general contractions of circular muscle (guinea-pig ileum) are consistently inhibited by prostaglandins E_1 and E_2 (BENNETT et al., 1968 a; FLESHLER and BENNETT, 1969). This response is not blocked by tetrodotoxin, hyoscine or pronethalol (BENNETT et al., 1968 a). There are significant qualitative differences in the response of longitudinal muscle strips to PGE and PGF (BERGSTRÖM et al., 1959; HORTON and MAIN, 1965). The response of the rabbit jejunum, for example, to $PGF_{2\alpha}$ is slower in onset and reaches its peak more slowly than the response to PGE_1 (Fig. 4) (HORTON and MAIN, 1965).

The responses of the rabbit duodenum to PGE_1 are reduced in size by the partial substitution of sodium in the bathing fluid by lithium. The contractions are unaffected by botulinum toxin in a

concentration which abolishes the response to nicotine suggesting that
the effects on this preparation are directly upon smooth muscle and
not mediated by cholinergic nerves (MIYAZAKI, ISHIZAWA, SUNANO,
SYUTO and SAKAGAMI, 1967). Contractions of the guinea-pig ileum to
PGE$_1$ are partially blocked by atropine (1—10 ng/ml) but the degree
of block cannot be increased by increasing the concentration of
atropine further (HORTON, 1965). Similar partial inhibition has been
observed with hyoscine and tetrodotoxin (Bennett et al., 1968 a). It

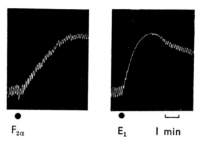

Fig. 4. Isotonic responses of isolated rabbit jejunum preparation suspended
in 4-ml organ-bath containing Tyrode solution. F$_{2\alpha}$=prostaglandin F$_{2\alpha}$,
    4 ng/ml; E$_1$ = prostaglandin E$_1$, 20 ng/ml (HORTON and MAIN, 1965)

seems likely that there is a small nervous component in the contractile
response of the guinea-pig ileum to PGE$_1$. Responses to PGE$_1$ are
unaffected by hexamethonium, methysergide or mepyramine (BEN-
NETT et al., 1968 a).

   Prostaglandins are released from the gastrointestinal tract (see
Chapter II). The possibility that they contribute to the motility of
intestinal smooth muscle in the peristaltic reflex (KOTTEGODA, 1969)
due to stimulation of non-cholinergic neurones in Auerbach's plexus
(AMBACHE and FREEMAN, 1968) has been considered. Other local mechanisms implicating prostaglandins have been proposed (BASS and
BENNETT, 1968).

3. Gastro-intestinal Motility Studied in vivo

   In their gastric secretion study, HORTON et al. (1968) experienced
subjective symptoms of increased gastrointestinal motility (mild colic
and diarrhoea) following oral administration of PGE$_1$ (10—40 µg/

kg). In specimens of gastric juice collected after PGE₁ administration there were copious amounts of bile — the appearance of the fluid resembled pea soup. The occurrence of bile in gastric fluid is an abnormal finding and suggests that massive regurgitation from the duodenum presumably through a relaxed pyloric sphincter had taken place.

More objective evidence of motility changes was obtained in a study on four volunteers in whom intestinal transit time was measur-

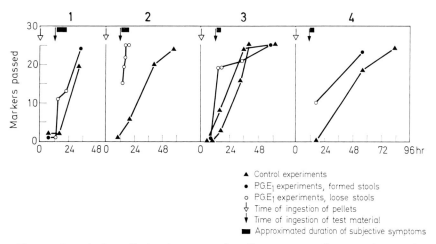

Fig. 5. Cumulative elimination-rate of radio-opaque pellets in the stools (MISIEWICZ et al., 1969)

ed by radio-opaque pellets and pressure measurements were made by a radio-telemetering capsule and by balloons inserted into the rectum and sigmoid colon. Oral administration of PGE₁ (2 mg) in these four subjects increased intestinal motility compared with control experiments (Fig. 5 and Table 2) (MISIEWICZ, WALLER, KILEY and HORTON, 1969). The most outstanding effect however was the passing of massive watery faeces 2 to 4 hours after the ingestion of PGE₁. The watery appearance of the faeces was reminiscent of cholera and it is of considerable interest that sodium transport across the gut is affected similarly by PGE₁ and the cholera toxin (GREENOUGH, PIERCE, AL AWQATI and CARPENTER, 1969). Cyclic AMP formation may be implicated as a common factor.

Table 2. *Number of progressive waves in left-colon pressure leads*

Volun- teer	Study	No. of progressive waves in experimental periods (15 min):										
		3	4	5	6	7	8	9	10	11	12	Total
1	Placebo	1	1	0	0	0	1	0	1	3	6	13
	Prostaglandin E_1	4	3	1	6	3	4	4	3	1	2	31
2	Placebo	1	0	0	0	2	0	1	1	0	0	5
	Prostaglandin E_1	0	0	0	0	0	0	0	0	9	0	9
3	Placebo	1	0	1	1	1	0	0	0	0	1	5
	Prostaglandin E_1	3	2	4	0	2	2	0	1	4	0	18
4	Placebo	0	2	0	1	0	0	1	0	0	0	4
	Prostaglandin E_1	2	1	2	4	0	2	1	2	2	4	20

(MISIEWICZ et al., 1969)

Increased intestinal motility *in vivo* has been reported in mice following intravenous PGE_1 (HOLMES and HORTON, 1968) and in the rat (BENNETT et al., 1968 b). Gastro-intestinal symptoms (vomiting and diarrhoea) have proved troublesome side-effects when using the higher doses of $PGF_{2\alpha}$ to terminate pregnancy.

References

AMBACHE, N., FREEMAN, M. A.: Atropine-resistant longitudinal muscle spasms due to excitation of non-cholinergic neurones in Auerbach's plexus. J. Physiol. (Lond.) **199**, 705—727 (1968).

BASS, P., BENNETT, D. R.: Local chemical regulation of motor action of the bowel — substance P and lipid-soluble acids. In: Handbook of Physiology Sec. 6, Alimentary Canal. Ed.: C. F. CODE. Washington: American Physiological Society 1968, pp. 2193—2212.

BENNETT, A., ELEY, K. G., SCHOLES, G. B.: Effects of prostaglandins E_1 and E_2 on human, guinea-pig and rat isolated small intestines. Brit. J. Pharmac. **34**, 630—638 (1968 a).

— — — Effect of prostaglandins E_1 and E_2 on intestinal motility in the guinea-pig and rat. Brit. J. Pharmac. **34**, 639—647 (1968 b).

BERGSTRÖM, S., ELIASSON, R., VON EULER, U. S., SJÖVALL, J.: Some biological effects of two crystalline prostaglandin factors. Acta physiol. scand. **45**, 133—144 (1959).

CLASSEN, M., KOCH, H., DEYHLE, P., WEIDENHILLER, S., DEMLING, L.: Wirkung von Prostaglandin E₁ auf die basale Magensekretion des Menschen. Klin. Wschr. **48**, 876—878 (1970).

FLESHLER, B., BENNETT, A.: Responses of human, guinea pig, and rat colonic circular muscle to prostaglandins. J. Lab. clin. Med. **74**, 872—873 (1969).

GHOSH, M. N., SCHILD, H. O.: Continuous recording of acid gastric secretion in the rat. Brit. J. Pharmac. **13**, 54—61 (1958).

GREENOUGH, W. B., PIERCE, N. F., AL AWQATI, Q., CARPENTER, C. C. J.: Stimulation of gut electrolyte secretion by prostaglandins, theophylline and cholera exotoxin. J. clin. Invest. **48**, 32 a—33 a (1969).

HORTON, E. W.: Biological activities of pure prostaglandins. Experientia **21**, 113—118 (1965).

— MAIN, I. H. M.: A comparison of the biological activities of four prostaglandins. Br. J. Pharmac. **21**, 182—189 (1963).

— — A comparison of the actions of prostaglandins F₂ₐ and E₁ on smooth muscle. Brit. J. Pharmac. **24**, 470—476 (1965).

— — THOMPSON, C. J., WRIGHT, P. M.: Effect of orally administered prostaglandin E₁ on gastric secretion and gastrointestinal motility in man. Gut **9**, 655—658 (1968).

JACOBSON, E. D.: Comparison of prostaglandin E₁ and norepinephrine on the gastric mucosal circulation. Proc. Soc. exp. Biol. Med. **133**, 516—519 (1970).

JONES, R. L.: Pharmacology of prostaglandins A and B. Ph.D. Thesis, University of London.

KOTTEGODA, S. R.: An analysis of possible nervous mechanisms involved in the peristaltic reflex. J. Physiol. (Lond.) **200**, 687—712 (1969).

MAIN, I. H. M.: Effects of prostaglandin E₂ (PGE₂) on the output of histamine and acid in rat gastric secretion induced by pentagastrin or histamine. Brit. J. Pharmac. **36**, 214—215P (1969).

MISIEWICZ, J. J., WALLER, S. L., KILEY, N., HORTON, E. W.: Effect of oral prostaglandin E₁ on intestinal transit in man. Lancet **1**, 648—651 (1969).

MIYAZAKI, E., ISHIZAWA, M., SUNANO, S., SYUTO, B., SAKAGAMI, T.: Stimulating action of prostaglandin on the rabbit duodenal muscle. Nobel Symposium 2, Prostaglandins. Eds.: S. BERGSTRÖM and B. SAMUELSSON. Stockholm: Almqvist and Wiksell 1967, pp. 277—281.

ROBERT, A.: Antisecretory property of prostaglandins. Prostaglandin Symposium of the Worcester Foundation for Exp. Biol. Eds.: P. W. RAMWELL and J. E. SHAW. New York: Interscience 1968, pp. 47—54.

— NEZAMIS, J. E., PHILLIPS, J. P.: Inhibition of gastric secretion by prostaglandins. Amer. J. dig. Dis. **12**, 1073—1076 (1967).

— — — Effect of prostaglandin E₁ on gastric secretion and ulcer formation in the rat. Gastroenterology **55**, 481—487 (1968).

— PHILLIPS, J. P., NEZAMIS, J. E.: Inhibition by prostaglandin E₁ of gastric secretion in the dog. Gastroenterology **54**, 1263 (1968).

RUDICK, J., GONDA, M., JANOWITZ, H. D.: Prostaglandin E_1: An inhibitor of electrolyte and stimulant of enzyme secretion in the pancreas. Fedn. Proc. Amer. Soc. exp. Biol. **29**, 445 Abs. (1970).

SHAW, J. E., RAMWELL, P. W.: Inhibition of gastric secretion in rats by prostaglandin E_1. Prostaglandin Symposium of the Worcester Foundation for Exp. Biol. Eds.: P. W. RAMWELL and J. E. SHAW. New York: Interscience 1968, pp. 55—66.

WAY, L., DURBIN, R. P.: Inhibition of gastric acid secretion *in vitro* by prostaglandin E_1. Nature **221**, 874—875 (1969).

WILSON, D. E., LEVINE, R. A.: Decreased canine gastric mucosal blood flow induced by prostaglandin E_1: A mechanism for its inhibitory effect on gastric secretion. Gastroenterology **56**, 1268 (1969).

XIV. Respiratory Tract Smooth Muscle

1. Isolated Preparations

The inhibitory action of PGE_1 on respiratory tract smooth muscle was first reported by MAIN in 1964. He studied isolated tracheal muscle preparations from seven species, monkey, cat, sheep, pig, guinea-pig, rabbit and ferret (Table 1). In those preparations with inherent tone a direct relaxant effect of PGE_1 is observed. In others, contractions of the muscle produced by the addition af acetylcholine, histamine, barium or dihydroergotamine, are inhibited by PGE_1. This can be demonstrated in one of two ways. If the PGE_1 is added to the organ bath before the stimulant drug, the contractile response is reduced compared with responses obtained with the same dose in the absence of PGE_1. If the PGE_1 is added after a sustained contraction has been obtained, the muscle relaxes (Fig. 1).

Table 1. *Threshold concentrations of prostaglandin E_1 causing inhibition of tone in isolated tracheal muscle preparations* (MAIN, 1964)

Dashes indicate that no inhibition could be demonstrated since the preparation had no initial tone even in the presence of the stimulant compound added. Values are µg/ml of prostaglandin E_1 which inhibited tone in the presence of the stimulants indicated

Species	Threshold concentration (µg/ml) of prostaglandin acting on Contraction due to				
	Acetyl-choline	Dihydro-ergotamine	Barium chloride	Histamine	Inherent tone
Cat	0.001	0.001	0.25	—	—
Monkey	0.02	—	> 2.5	—	—
Rabbit	0.05	—	—	—	—
Guinea-pig	0.005	—	< 0.25	0.005	0.005
Ferret	0.005	—	0.25	—	0.005
Sheep	3.0	—	0.05	—	> 1.0
Pig	0.25	—	—	—	> 2.0

The isolated tracheal muscle of the cat is the most sensitive preparation, often responding to a PGE_1 concentration of 1 ng/ml. Tracheal muscle from the monkey is almost as sensitive (Fig. 2 c). The rabbit trachea is less sensitive and complete relaxation is difficult to obtain even on repeated dosage with PGE_1.

The inherent tone of guinea-pig and ferret tracheal preparations is reduced by PGE_1 as are the contractions to acetylcholine (Fig. 2). Pig tracheal muscle contractions to acetylcholine are reduced by PGE_1 but sheep muscle is very insensitive.

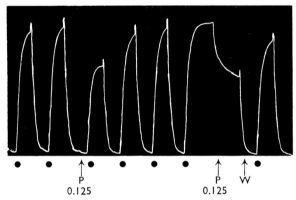

Fig. 1. Responses of a cat isolated trachea preparation, suspended in a 4 ml organ-bath containing Krebs-Henseleit solution. At the dots acetylcholine (0.5 µg/ml) was added. P = prostaglandin E_1 (0.125 µg/ml); W = wash (MAIN, 1964)

PGE_2, PGE_3, $PGF_{1\alpha}$ and $PGF_{2\alpha}$ all antagonise acetylcholine on the cat isolated tracheal preparation (Fig. 3). Their activities relative to PGE_1 ($=1.0$) were estimated to be 1.0, 0.2, 0.002 and 0.03 respectively (MAIN, 1964; HORTON and MAIN, 1965).

Of the seven preparations which MAIN tested, only the guinea-pig trachea contracted in response to histamine and only the cat responded to dihydroergotamine (Fig. 4). These responses were antagonised by low concentrations of PGE_1. Barium chloride (0.1—1.0 mg/ml)-induced contractions of tracheal muscle from cat, guinea-pig (Fig. 5), ferret and sheep were inhibited by PGE_1 but in five experiments barium-induced contractions of monkey tracheal muscle were unaffected by PGE_1 (0.25 to 5 µg/ml). PGE_1 also relaxes tracheal smooth muscle of the dog (TÜRKER and KHAIRALLAH, 1969).

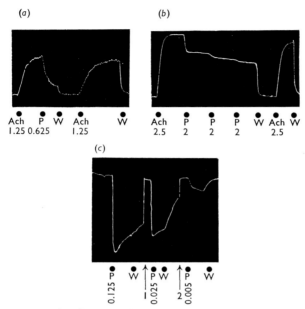

Fig. 2. Responses of isolated trachea preparations suspended in 4 ml organ-baths containing Krebs-Henseleit solution. (a) monkey, (b) rabbit and (c) ferret. Ach = acetylcholine; P = prostaglandin E_1; W = wash. All drug concentrations are in µg/ml. At arrows 1 and 2, the drum was stopped for 20 and 10 min respectively (MAIN, 1964)

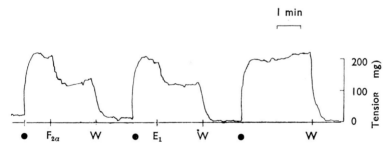

Fig. 3. Isometric responses of cat isolated trachea, suspended in 4-ml organ-bath containing Krebs-Henseleit solution. At the dots acetylcholine (12.5 ng/ml) was added. $F_{2\alpha}$ = prostaglandin $F_{2\alpha}$, 0.75 µg/ml; E_1 = prostaglandin E_1, 1.9 ng/ml; W = wash (HORTON and MAIN, 1965)

The response of human isolated bronchial smooth muscle to pro-staglandins has been investigated (COLLIER and SWEATMAN, 1968; SWEATMAN and COLLIER, 1968; SHEARD, 1968). PGE_1 and PGE_2

relax this preparation but $PGF_{2\alpha}$ (0.8—800 ng/ml) contracts it. This stimulant action of $PGF_{2\alpha}$ is reduced by simultaneous administration of PGE_1 or PGE_2 but not by either atropine (1 µg/ml) or mepyramine (1 µg/ml). Furthermore although tachyphylaxis to SRS-A and $PGF_{2\alpha}$ could be produced there was no cross tachyphylaxis between these substances. Non-steroidal anti-inflammatory drugs such as meclofenamate (100 ng/ml) reduced the contractile responses of human bronchial smooth muscle to $PGF_{2\alpha}$ but not to acetylcholine (Fig. 6).

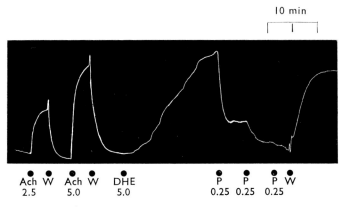

Fig. 4. Responses of cat isolated trachea preparation, suspended in a 4 ml organ-bath containing Krebs-Henseleit solution. Ach = acetylcholine; DHE = dihydroergotamine; P = prostaglandin E_1; W = wash. All drug concentrations are in µg/ml (MAIN, 1964)

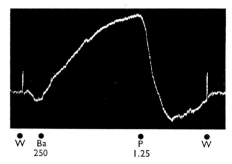

Fig. 5. Responses of a guinea-pig isolated trachea preparation, suspended in a 4 ml organ-bath containing Krebs-Henseleit solution. Ba = barium chloride; P = prostaglandin E_1; W = wash. Both drug concentrations are in µg/ml (MAIN, 1964)

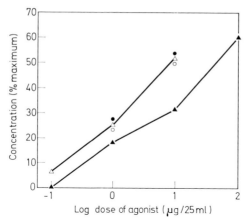

Fig. 6. Effect of meclofenamate on the dose response lines of prostaglandin $F_{2\alpha}$ ($PGF_{2\alpha}$) on human isolated bronchial muscle. ($\triangle$) $PGF_{2\alpha}$ without meclofenamate; ($\blacktriangle$) $PGF_{2\alpha}$, in the presence of 100 ng/ml of meclofenamate; ($\bigcirc$) acetylcholine without meclofenamate; ($\bullet$) acetylcholine in the presence of 100 ng/ml of meclofenamate (COLLIER and SWEATMAN, 1968)

2. In vivo Experiments

The bronchoconstrictor response to vagal stimulation or to histamine in guinea-pigs is reduced and sometimes temporarily abolished by PGE_1 (0.1 µg/kg) or $PGF_{2\alpha}$ (0.5 µg/kg) injected intravenously (Fig. 7) (MAIN, 1964). Similar antagonism of bronchoconstriction has been observed in the rabbit with PGE_1 (Fig. 8). In some preparations a decrease in bronchial resistance after PGE_1 can be demon-

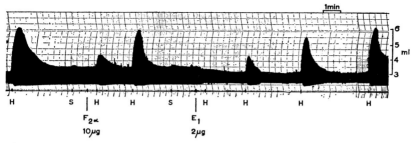

Fig. 7. Guinea-pig (400 g) anaesthetised with urethane open chested and artifically ventilated. Tracing of tidal overflow volume using the Konzett and Rössler technique. At H 0.5 µg histamine i.v. S = 0.2 ml saline. Responses to $F_{2\alpha}$ (10 µg) and E_1 (2 µg) are shown Experiment by I. H. M. MAIN)

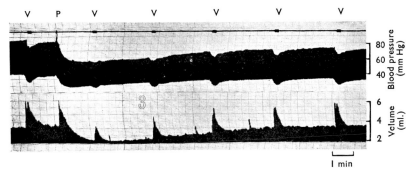

Fig. 8. Rabbit, 2.5 kg, anaesthetized with urethane. Uppermost trace, event marker; middle trace, arterial blood pressure; lowest trace, tidal overflow volume. V = stimulation of the left vagus nerve for 10 sec; P = prostaglandin E₁ (4 µg) injected intravenously (MAIN, 1964)

strated without first inducing bronchoconstriction by nerve stimulation or histamine.

This bronchodilator action of PGE₁ *in vivo* has been confirmed in the guinea-pig by LARGE, LESWELL and MAXWELL (1969) who showed that PGE₁ in an aerosol is particularly effective in antagonising the bronchoconstrictor response to histamine. On intravenous injection PGE₁ is less active than isoprenaline but by aerosol administration it is some 70 times more active (Table 2). ADOLPHSON and TOWNLEY (1970) have confirmed these results in unanaesthetised guinea-pigs.

Table 2. *Resistance to inflation by recording tidal volume at constant inflation pressure or changes in intra-tracheal pressure on inflation at constant volume*

Intravenous				Aerosol (for 20 inflations)			
Isoprenaline		PGE₁		Isoprenaline		PGE₁	
Dose (µg)	Percentage inhibition bronchoconstriction	Dose	Percentage inhibition	Concentration (µg/ml)	Percentage inhibition	Concentration	Percentage inhibition
0.05	42 (2)	0.05	20 (1)	100	21 ± 4.1 (8)	5	22 (1)
0.1	67 (2)	0.1	35 (2)	500	25 (1)	10	53 ± 7.2 (5)
0.2	86 (1)	0.2	58 (2)	1000	57 ± 5.7 (11)	50	55 ± 11.7 (4)
		0.3	76 (1)			100	67 ± 7.1 (10)

(LARGE, LESWELL and MAXWELL, 1969)

In contrast both PGE_1 (0.3 µg/kg) and $PGF_2\alpha$ (15 µg/kg) injected intravenously in the cat increase bronchial resistance as measured by the KONZETT and RÖSSLER technique (MAIN, 1964; ÄNGGÅRD and BERGSTRÖM, 1963) (Fig. 9). A similar response to large doses of $PGF_{2\alpha}$

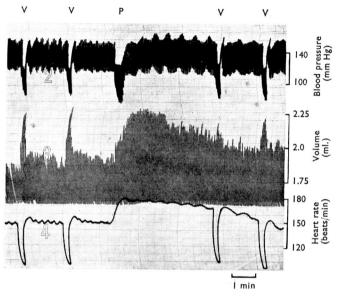

Fig. 9. Cat, 2.8 kg, anaesthetized with chloralose and urethane. Uppermost trace, arterial blood pressure; middle trace, tidal overflow volume; lowest trace, heart rate. V = stimulation of left vagus nerve for 10 sec; P = prostaglandin E_1, 10 µg injected intravenously (MAIN, 1964)

has been obtained *in vivo* in the guinea-pig (BERRY and COLLIER, 1964) and this response unlike that to bradykinin and SRS-A is not blocked by calcium acetylsalicylate. The possibility that actions on pulmonary vessels may contribute to these effects must be considered (SAID, 1968).

3. Human Volunteers and Asthmatics

A bronchodilator action of PGE_1 in human asthmatic subjects has been reported, suggesting a possible clinical application (CUTHBERT, 1969). Forced expiratory volume in 1 second (FEV_1) was measured in healthy and asthmatic subjects before and after the administration of either PGE_1 or isoprenaline by metered aerosol. It was observed that the free acid of PGE_1 was irritant to the respiratory tract but the triethanolamine salt was free from irritant activity.

In the healthy volunteers, FEV_1 was unchanged by PGE_1. In 5 asthmatic subjects PGE_1 (triethanolamine) 55 mg was rather more effective than isoprenaline 550 mg. In one subject FEV_1 was reduced following PGE_1.

The place of PGE_1 in the treatment of bronchial asthma or other obstructive disorders of the respiratory tract has yet to be established. The results from these preliminary investigations are however encouraging.

References

ADOLPHSON, R. L., TOWNLEY, R. G.: A comparison of the bronchodilator activities of isoproterenol and prostaglandin E_1 aerosols. J. Allergy **45**, 119—120 (1970).

ÄNGGÅRD, E., BERGSTRÖM, S.: Biological effects of an unsaturated trihydroxy acid ($PGF_{2\alpha}$) from normal swine lung. Acta physiol. scand. **58**, 1—12 (1963).

BERRY, P. A., COLLIER, H. O. J.: Bronchoconstrictor action and antagonism of a slow-reacting substance from anaphylaxis of guinea-pig isolated lung. Brit. J. Pharmac. **23**, 201—216 (1964).

COLLIER, H. O. J., SWEATMAN, W. J. F.: Antagonism by fenamates of prostaglandin $F_{2\alpha}$ and of slow reacting substance on human bronchial muscle. Nature (Lond.) **219**, 864—865 (1968).

CUTHBERT, M. F.: Effect on airways resistance of prostaglandin E_1 given by aerosol to healthy and asthmatic volunteers. Brit. med. J. **4**, 723—726 (1969).

HORTON, E. W., MAIN, I. H. M.: A comparison of the actions of prostaglandins $F_{2\alpha}$ and E_1 on smooth muscle. Brit. J. Pharmac. **24**, 470—476 (1965).

LARGE, B. J., LESWELL, P. F., MAXWELL, D. R.: Bronchodilator activity of an aerosol of prostaglandin E_1 in experimental animals. Nature (Lond.) **224**, 78—80 (1969).

MAIN, I. H. M.: The inhibitory actions of prostaglandins on respiratory smooth muscle. Brit. J. Pharmac. **22**, 511—519 (1964).

SAID, S. I.: Some respiratory effects of prostaglandins E_2 and $F_{2\alpha}$. Prostaglandin Symposium of the Worcester Foundation for Exp. Biol. Ed.: P. W. RAMWELL and J. E. SHAW. New York: Interscience 1968, pp. 267—277.

SHEARD, P.: The effect of prostaglandin E_1 on isolated bronchial muscle from man. J. Pharm. Pharmac. **20**, 232—233 (1968).

SWEATMAN, W. J. F., COLLIER, H. O. J.: Effects of prostaglandins on human bronchial muscle. Nature (Lond.) **217**, 69 (1968).

TÜRKER, R., KHAIRALLAH, P. A.: Prostaglandin E_1 action on canine isolated tracheal muscle. J. Pharm. Pharmac. **21**, 498—501 (1969).

XV. Biological Significance of the Prostaglandins

1. Reproduction

It has long been suspected that the high concentrations of pro-staglandins in human semen are essential for normal fertility (ASPLUND, 1947; HAWKINS and LABRUM, 1961). Evidence which strongly supports this contention has recently been reported by BYG-DEMAN, FREDRICSSON, SVANBORG and SAMUELSSON (1970). The concentrations of different prostaglandins in human seminal fluid were estimated by a physicochemical method (U.V. absorption at 278 nm after treatment with alkali). The seminal PGE concentration of 21 men in infertile marriages with no abnormal clinical or laboratory findings was lower (18.1 µg/ml) than that of men of proven recent fertility (54.4 µg/ml). This difference is statistically highly significant (P < 0.001). There was no difference however in the concentrations of other prostaglandins (PGA, PGB and their 19-hydroxy derivatives) between the two groups. It seems likely that infertile couples in which the husband's semen is deficient in PGE could be treated by replacement therapy. It is therefore desirable in future that PGE determinations should be included routinely in the investigation of infertility.

The mechanism by which PGE compounds aid conception is unknown. Washed human sperm suspended in artificial plasma show no difference in motility whether PGE_1 is added or not (HORTON, 1965). The presence of PGE_1, 100 µg/ml, also had no effect on sperm oxygen uptake, the amount of fructose metabolised or the amount of lactic acid accumulating (ELIASSON, MURDOCH and WHITE, 1968), though both PGE_1 and $PGF_{1\alpha}$ increase oxidation of glucose to carbon dioxide by ram epididymal sperm (PENTO, CENEDELLA and INSKEEP, 1970).

Smooth muscle contractions of ejaculation, in particular the emptying of the seminal vesicles, may be stimulated by seminal pro-staglandins. In the guinea-pig, contractile responses of the seminal

12*

vesicles and vas deferens to catecholamine and hypogastric nerve stimulation are enhanced by PGE_1 (ELIASSON and RISLEY, 1966; MANTEGAZZA and NAIMZADA, 1965). Similar studies have not yet been reported with human tissue.

Prostaglandins affect female reproductive tract smooth muscle (Chapter VI). When deposited in the posterior fornix of the vagina at coitus, they may act locally on the cervix or the body of the uterus or they may be absorbed from the vagina into the systemic circulation reaching target organs including the Fallopian tubes. The production of biologically active substances by one individual and their transfer to and action upon the target organs in a second individual is an interesting biological mechanism (HORTON, MAIN and THOMPSON, 1963). If it is shown to occur physiologically in man, seminal prostaglandins must be classified as a special category of pheromones.

Prostaglandins by acting upon the female reproductive tract may affect sperm transport or by contracting the uterine end of the Fallopian tube whilst inhibiting contractions of the more distal segments, they may allow access of the sperm to the ovum but delay transport of the ovum into the uterine cavity until fertilisation and adequate blastocyst development have taken place (SANDBERG, INGELMAN-SUNDBERG and RYDÉN, 1963). Mice receiving PGE_1 1 mg/kg on the day of mating and on the next day had 22% more foetuses per pregnant mouse than saline-treated controls (HORTON and MARLEY, 1969).

There is evidence for the presence of an oxytocic lipid in blood at the time of menstruation (PICKLES, 1959). Moreover a substance which behaves chromatographically and biologically like $PGF_{2\alpha}$ has been detected in the circulation immediately before and during the contractions of labour (KARIM, 1968). PGE_2 and $PGF_{2\alpha}$ have been identified in menstrual fluid, human endometrium, amniotic fluid and the decidua (EGLINTON, RAPHAEL, SMITH, HALL and PICKLES, 1963; KARIM, 1966; KARIM and DEVLIN, 1967). There is sufficient evidence to suggest as a working hypothesis that one or more prostaglandins may be implicated in menstruation, in some cases of spontaneous abortion and in the uterine contractions of parturition (KARIM, 1969).

Prostaglandins constrict placental blood vessels (VON EULER, 1938; HILLIER and KARIM, 1968) and occur in high concentrations in human umbilical cord (KARIM, 1967). Their release at birth may contribute to constriction of umbilical vessels, and their possible implication in pathological changes in placental blood flow during pregnancy merits investigation.

2. Prostaglandins as Circulating Hormones

Prostaglandins are released into the circulation from various organs on chemical or nerve stimulation, but a very high proportion of PGE and PGF compounds are removed on one circulation through the lungs. This may be regarded as evidence against the hypothesis that these compounds function as hormones in the classical sense. This argument does not apply to PGA compounds which are not taken up to any great extent by the lungs and so could reach a target organ. No hormonal function can at present be assigned to the PGA compounds; perhaps the most likely role is that of a natriuretic hormone (see Chapter XI).

$PGF_{2\alpha}$ does reach the uterine venous blood under physiological conditions. It is probably identical with the hormone luteolysin (POYSER, HORTON, THOMPSON and LOS, 1970, 1971; BLATCHLEY, DONOVAN, POYSER, HORTON, THOMPSON and LOS, 1971; BLAND, HORTON and POYSER, 1971; HORTON, JONES, POYSER and THOMPSON, 1971). $PGF_{2\alpha}$ released from the uterus into the uterine venous blood can reach the ipsilateral ovary via a counter-current mechanism between the utero-ovarian vein and the ovarian artery in the sheep (McCRACKEN, 1971).

3. Interaction between Prostaglandins and Catecholamines

VON EULER (1939) noted that prostaglandin injected intravenously reduced the size of pressor response to adrenaline in the rabbit. This observation was confirmed by HOLMES, HORTON and MAIN (1963) using pure PGE_1 and they showed that other pressor and vasoconstrictor substances (vasopressin, angiotensin and noradrenaline) were also antagonised by PGE_1 (Fig. 1). About the same time STEINBERG, VAUGHAN, NESTEL and BERGSTRÖM (1963) reported that PGE_1 blocks the lipolytic action of adrenaline and other lipolytic hormones. Both groups concluded that PGE_1 acts upon a site on the biochemical pathway which is activated by several different hormones. In the case of adipose tissue this proved to be adenyl cyclase (see below). Whether a similar mechanism can account for the antagonism on smooth muscle is unknown. The antagonism by PGE_1 of noradrenaline responses of the cerebellar Purkinje cells almost certainly involve adenyl cyclase (see Chapter IX and below).

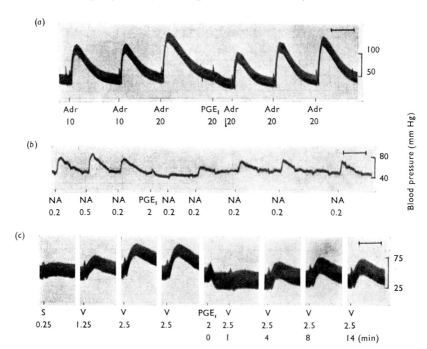

Fig. 1. Records of carotid arterial blood pressure from: (a) a rabbit (2 kg) anaesthetized with urethane; (b) a rat (290 g) anaesthetized with urethane; and (c) a pithed rat (200 g). Drugs were injected intravenously. Adr = adrenaline (µg); NA = noradrenaline (µg); V = vasopressin (mU); PGE_1 = prostaglandin E_1 (µg); S = saline (ml). Time calibrations are 1 min. In (c) the lower figures indicate times in minutes after administration of prostaglandin E_1 at which doses of vasopressin were given (HOLMES et al., 1963)

CLEGG (1966) studied the interaction of prostaglandins and sympathomimetic amines on several isolated smooth muscle preparations. She concluded that PGE_1 and $PGF_{2\alpha}$ antagonise sympathomimetic amines on these tissues irrespective of the response produced by the amine (stimulation or inhibition) and irrespective of the nature of the direct action to the prostaglandin (stimulation, inhibition or no effect) on the tissue. The antagonism in some tissues (rat fundus, rat vas deferens and guinea-pig seminal vesicles) is preceded by potentiation. With the rabbit trachea, potentiation was the only effect observed. The initial potentiation occurs both in preparations which are inhibited and in preparations which are contracted by

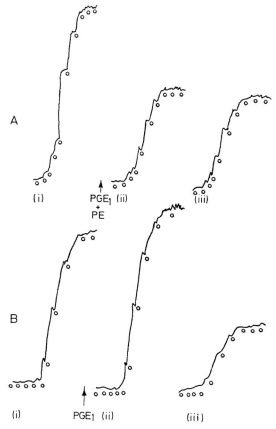

Fig. 2. Tracings of cumulative dose response curves of rat vas deferens stimulated by phenylephrine. The concentration of phenylephrine was increased from 10^{-7} M to 3×10^{-3} M. The open circles represent the points at which phenylephrine was added. Both preparations were exposed to PGE_1 (4×10^{-7} g/ml) for 5 min, one in the presence of 10^{-2} M phenylephrine (A), one in the absence of phenylephrine (B), Curves A_{ii} and B_{ii} were made 100 min after exposure to PGE_1, curves A_{iii} and B_{iii} 160 min after exposure to PGE_1 (CLEGG, 1966)

sympathomimetics, and occurs irrespective of the nature of the direct effect of the prostaglandin. Thus the inhibitory responses of the rat uterus to adrenaline and of the rat fundus to noradrenaline are initially potentiated by prostaglandins although both PGE_1 and $PGF_{2\alpha}$ have a direct stimulant effect on both tissues.

Using the rat vas deferens preparation, CLEGG, 1966 showed that exposure to a high concentration of agonist (phenylephrine) did not reduce the subsequent inhibition of responses to the agonist by PGE_1. Indeed inhibition appeared more rapidly and to a greater extent (Fig. 2). One interpretation of this result is that PGE_1 increases the affinity of the sympathomimetic amines to the adrenergic receptor site. This could explain the initial potentiation followed by a depression of the response as the number of available receptor sites is reduced. If this is a physiological mechanism, prostaglandins must presumably be implicated in the modulation of smooth muscle responsiveness to chemical mediators.

In several tissues PGE release has been detected on sympathetic nerve stimulation. For example on stimulating the splenic nerve to the blood perfused spleen of the dog, PGE_2 and $PGF_{2\alpha}$ are released into the venous effluent (DAVIES, HORTON and WITHRINGTON, 1967; GILMORE, VANE and WYLLIE, 1968). The identification of these prostaglandins has now been confirmed by combined gas chromatography and mass spectrometry. The significance of this release is, however, unknown since prostaglandins have no substantial effects on the dog spleen nor do they modify its response to nerve stimulation or catecholamines (DAVIES and WITHRINGTON, 1968).

In the cat, however, PGE compounds reduce the amount of noradrenaline released from the spleen on nerve stimulation (HEDQVIST, 1968). Hedqvist has extended these observations to other sites and suggests that prostaglandins released from post-synaptic sites in response to the transmitter, function as modulators reducing the output of transmitter to continued nerve stimulation (see Chapter IX).

4. Prostaglandins and Adenyl Cyclase

STEINBERG and his co-workers (1963) made the original observation that PGE_1 inhibits hormone-induced lipolysis. Hormones stimulate lipolysis via adenyl cyclase and the formation of cyclic AMP.

Since the lipolytic action of cyclic AMP itself it *not* blocked by PGE$_1$ and since PGE$_1$ inhibits the accumulation of cyclic AMP in response to noradrenaline, it is concluded that PGE$_1$ inhibits the enzyme adenyl cyclase (Fig. 3). Furthermore, since nerve and hormone stimulation of adipose tissue release prostaglandins in amounts

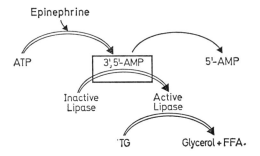

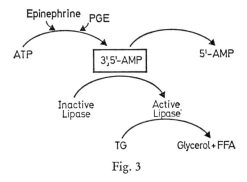

Fig. 3

sufficient to affect lipolysis (SHAW and RAMWELL, 1968), these endogenous prostaglandins may regulate the activity of the lipolytic hormones. An intact cell membrane is required for this action of PGE$_1$. In rat fat cell particles adenyl cyclase can be activated by adrenaline but PGE$_1$ has no inhibitory action. Therefore, PGE$_1$ appears to act on the adenyl cyclase system by an indirect action (see below) (RAMWELL and SHAW, 1970).

A similar mechanism may account for the inter-action of PGE$_1$ with hormones on other tissues. Vasopressin increases water permeability of the isolated toad bladder and rabbit renal tubule via adenyl cyclase and cyclic AMP. PGE_1 blocks the action of vasopressin possibly by a mechanism analogous to that proposed for adipose tissue. Similarly, gastric secretion of acid response to histamine or pentagastrin is mediated via cyclic AMP and inhibited by PGE_1. The antagonism between noradrenaline and PGE compounds on cerebellar Purkinje cells referred to above is similarly mediated.

On the other hand, there is abundant evidence that on many tissues PGE compounds, although affecting adenyl cyclase, do not inhibit the stimulant hormone but mimic its action. Thus PGE_2 like corticotrophin stimulates corticosteroidogenesis (FLACK et al., 1969), an effect which is mediated by cyclic AMP. The action of both stimulants is reduced by cycloheximide. PGE_1 mimicks the action of thyroid stimulating hormone on the thyroid, both acting by increasing cyclic AMP formation (KANEKO, ZOR and FIELD, 1969) (see Chapter VII). On the beta cells of the islets of Langerhans, glucose, cyclic AMP and PGE_1 all stimulate insulin secretion, the effect of PGE_1 is probably mediated here by cyclic AMP also. Platelet aggregation is known to be associated with a decrease in cyclic AMP levels. Again PGE_1 probably exerts its inhibitory action by increasing cyclic AMP concentrations. Finally on the bovine corpus luteum PGE_2 stimulates steroidogenesis (like luteinizing hormone). However results are not always so clear cut, for example BEDWANI and HORTON (1968) using rabbit ovaries found that PGE_2 increased levels of 20α-dihydroprogesterone but had no effect on progesterone itself; moreover PGE_1 had no effect on the biosynthesis of either steroid.

It may be concluded that interactions between PGE compounds and adenyl cyclase may underlie many of the effects produced by prostaglandins, but as pointed out previously (HORTON, 1969) the hypothesis does nothing to explain the mode of action of PGF or PGA compounds, though their pharmacological activity is often greater and at least as likely to reflect a physiological role as that of the PGE compounds. Furthermore, the hypothesis does not account satisfactorily for the many apparent differences between PGE_1 and PGE_2 which tend to be forgotten. We know little about the mechanism which determines which prostaglandin will be released at a particular site in response to a particular stimulus and most of the evidence of identification of these minute amounts is so inadequate that quantitative data so far obtained are necessarily of doubtful value.

5. Prostaglandins and Ions

Adenyl cyclase activity in rat fat cell particles or ghosts, in adrenal mitochondria, adrenal tumour cells and bone is decreased by an increase in calcium ion concentration. Conversely, in fat cells the adenyl cyclase activity is increased by an increase in sodium ion concentration (see RAMWELL and SHAW, 1970 for references). Thus if PGE_1 displaces membrane bound calcium and causes an influx of sodium (as it does in frog skin), activation of adenyl cyclase will be favoured. Experiments on human and turkey erythrocytes believed to lack adenyl cyclase, indicate that PGE_1 causes an influx of sodium in the absence of this enzyme. Thus the primary response to PGE_1 appears to be a change in membrane permeability to ions which secondarily activate adenyl cyclase. RAMWELL and SHAW (1970) suggest that displacement of membrane calcium may be the fundamental mechanism as postulated earlier by PICKLES and by WOLFE and COCEANI (PICKLES, HALL, CLEGG and SULLIVAN, 1966; COCEANI and WOLFE, 1966; COCEANI, DREIFUSS, PUGLISI and WOLFE, 1969).

References

ASPLUND, J.: A quantitative determination of the content of contractive substances in human sperm and their significance for the motility and vitality of the spermatozoa. Acta physiol. scand. 13, 103—108 (1947).

BEDWANI, J. R., HORTON, E. W.: The effects of prostaglandins E_1 and E_2 on ovarian steroidogenesis. Life Sci. 7, 389—393 (1968).

BLAND, K. P., HORTON, E. W., POYSER, N. L.: Levels of prostaglandin $F_{2\alpha}$ in uterine venous blood of sheep during the oestrous cycle. Life Sci. 10, 509—517 (1971).

BLATCHLEY, F. R., DONOVAN, B. T., POYSER, N. L., HORTON, E. W., THOMPSON, C. J., LOS, M.: Identification of prostaglandin $F_{2\alpha}$ in the utero-ovarian blood of guinea-pig after treatment with oestrogen. Nature (Lond.) 230, 243—244 (1971).

BYGDEMAN, M., FREDRICSSON, B., SVANBORG, K., SAMUELSSON, B.: The relation between fertility and prostaglandin content of seminal fluid in man. Fert. Steril. 21, 622—629 (1970).

CLEGG, P. C.: The effect of prostaglandins on the response of isolated smooth-muscle preparations to sympathomimetic substances. Mem. Soc. Endocr. 14, 119—136 (1966).

COCEANI, F., DREIFUSS, J. J., PUGLISI, L., WOLFE, L. S.: Prostaglandins and membrane function. In: Prostaglandins, peptides and amines. Ed.: P. MANTEGAZZA and E. W. HORTON. London: Academic Press 1969, pp. 73—84.

— WOLFE, L. S.: On the action of prostaglandin E_1 and prostaglandins from brain on the isolated rat stomach. Can. J. Physiol. Pharmac. **44**, 933—950 (1966).

DAVIES, B. N., HORTON, E. W., WITHRINGTON, P. G.: The occurrence of prostaglandin E_2 in splenic venous blood of the dog following nerve stimulation. J. Physiol. (Lond.) **188**, 38—39P (1967).

— WITHRINGTON, P. G.: The effects of prostaglandin E_1 and E_2 on the smooth muscle of the dog spleen and on its responses to catecholamines, angiotensin and nerve stimulation. Brit. J. Pharmac. **32**, 136—144 (1968).

EGLINTON, G., RAPHAEL, R. A., SMITH, G. N., HALL, W. J., PICKLES, V. R.: The isolation and identification of two smooth muscle stimulants from menstrual fluid. Nature (Lond.) **200**, 960, 993—995 (1963).

ELIASSON, R., MURDOCH, R. N., WHITE, I. G.: The metabolism of human spermatozoa in the presence of prostaglandin E_1. Acta physiol. scand. **73**, 379—382 (1968).

— RISLEY, P. L.: Potentiated response of isolated seminal vesicles to catecholamines and acetylcholine in the presence of PGE_1. Acta physiol. scand. **67**, 253—254 (1966).

FLACK, J. D., JESSUP, R., RAMWELL, P. W.: Prostaglandin stimulation of rat corticosteroidogenesis. Science N. Y. **163**, 691—692 (1969).

GILMORE, N., VANE, J. R., WYLLIE, J. H.: Prostaglandin output from the spleen. Brit. J. Pharmac. **32**, 425—426P (1968).

HAWKINS, D. F., LABRUM, A. H.: Semen prostaglandin levels in fifty patients attending a fertility clinic. J. Reprod. Fert. **2**, 1—10 (1961).

HEDQVIST, P.: Reduced effector response to nerve stimulation in the cat spleen after administration of prostaglandin E_1. Acta physiol. scand. **74**, 7A (1968).

HILLIER, K., KARIM, S. M. M.: Effects of prostaglandins E_1, E_2, $F_{1\alpha}$ and $F_{2\alpha}$ on isolated human umbilical and placental blood vessels. J. Obstet. Gynaec. Brit. Commonw. **75**, 667—673 (1968).

HOLMES, S. W., HORTON, E. W., MAIN, I. H. M.: The effect of prostaglandin E_1 on responses of smooth muscle to catecholamines, angiotensin and vasopressin. Brit. J. Pharmac. **21**, 538—543 (1963).

HORTON, E. W.: Biological activities of pure prostaglandins. Experientia **21**, 113—118 (1965).

— Hypotheses on physiological roles of prostaglandins. Physiol. Rev. **49**, 122—161 (1969).

— JONES, R. L., POYSER, N. L., THOMPSON, C. J.: The release of prostaglandins. Ann. N. Y. Acad. Sci. **180**, 351—361 (1971).

— MAIN, I. H. M., THOMPSON, C. J.: The action of intravaginal prostaglandin E₁ on the female reproductive tract. J. Physiol. (Lond.) **168**, 54—55P (1963).

— MARLEY, P. B.: An investigation of the possible effects of prostaglandins E₁, F₂α and F₂β on pregnancy in mice and rabbits. Brit. J. Pharmac. **36**, 188P (1969).

KARIM, S. M. M.: Identification of prostaglandins in human amniotic fluid. J. Obstet. Gynaec. Brit. Commonw. **73**, 903—908 (1966).

— The identification of prostaglandins in human umbilical cord. Brit. J. Pharmac. **29**, 230—237 (1967).

— Appearance of prostaglandin F₂α in human blood during labour. Brit. med. J. **4**, 618—621 (1968).

— The role of prostaglandin F₂α in parturition. In: Prostaglandins, Peptides and Amines. Ed.: P. MANTEGAZZA and E. W. HORTON. London: Academic Press 1969, pp. 65—72.

— DEVLIN, J.: Prostaglandin content of amniotic fluid during pregnancy and labour. J. Obstet. Gynaec. Brit. Commonw. **74**, 230—234 (1967).

McCRACKEN, J. A.: Prostaglandin F₂α and corpus luteum regression. Ann. N. Y. Acad. Sci. **180**, 456—469 (1971).

MANTEGAZZA, P., NAIMZADA, M.: Attività della prostaglandina E₁ sul preparato nervo ipogastrico-deferente di varie specie animali. Atti Accad. med. lomb. **20**, 58—64 (1965).

PENTO, J. T., CENEDELLA, R. J., INSKEEP, E. K.: Effects of prostaglandins E₁ and F₁α upon carbohydrate metabolism of ejaculated and epididymal ram spermatozoa *in vitro*. J. Anim. Sci. **30**, 409—411 (1970).

PICKLES, V. R.: Some evidence that the human endometrium produces a hormone that stimulates plain muscle. J. Endocr. **18**, i—ii (1959).

— HALL, W. J., CLEGG, P. C., SULLIVAN, T. J.: Some experiments on the mechanism of action of prostaglandins on the guinea-pig and rat myometrium. Mem. Soc. Endocr. **14**, 89—103 (1966).

POYSER, N. L., HORTON, E. W., THOMPSON, C. J., LOS, M.: Identification of prostaglandin F₂α released by distension of the guinea-pig uterus in vitro. J. Endocrin. **48**, xliii (1970).

— — — — Identification of prostaglandin F₂α released by distension of the guinea-pig uterus in vitro. Nature **230**, 526—528 (1971).

RAMWELL, P. W., SHAW, J. E.: Biological significance of the prostaglandins. Rec. Progr. Hormone Res. **26**, 139—187 (1970).

SANDBERG, F., INGELMAN-SUNDBERG, A., RYDÉN, G.: The specific effect of prostaglandin on different parts of the human fallopian tube. J. Obstet. Gynaec. Brit. Commonw. **70**, 130—134 (1963).

SHAW, J. E., RAMWELL, P. W.: Prostaglandin release from the adrenal gland. Nobel Symposium 2, Prostaglandins. Ed.: S. BERGSTRÖM and B. SAMUELSSON. Stockholm: Almqvist and Wiksell 1967, pp. 293—299.

— — Release of prostaglandin from rat epididymal fad pad on nervous and hormonal stimulation. J. biol. Chem. **243**, 1498—1503 (1968).

STEINBERG, D., VAUGHAN, M., NESTEL, P., BERGSTRÖM, S.: Effects of prostaglandin E opposing those of catecholamines on blood pressure and on
tri-glyceride breakdown in adipose tissue. Biochem. Pharmac. **12,**
764—766 (1963).

VON EULER, U. S.: Action of adrenaline, acetylcholine and other substances
on nerve-free vessels (human placenta). J. Physiol. (Lond.) **93,** 129—143
(1938).

— Weitere Untersuchungen über Prostaglandin, die physiologisch aktive
Substanz gewisser Genitaldrüsen. Skand. Arch. Physiol. **81,** 65—80
(1939).

Subject Index

Monographs on Endocrinology

Already Published:

Vol. 1: OHNO, S., Sex Chromosomes and Sex-linked Genes. With 33 figures. X, 192 pages. 1967. DM 38,—

Vol. 2: EIK-NES, K. B., and E. C. HORNING, Gas Phase Chromatography of Steroids. With 85 figures. XV, 382 pages. 1968. DM 38,—

Vol. 3: SULMAN, F. G., Hypothalamic Control of Lactation. With 58 figures. XII, 235 pages. 1970. DM 52,—

Vol. 4: WESTPHAL, U., Steroid-Protein Interactions. With 144 figures. XIII, 567 pages. 1971. DM 86,—

Vol. 5: MÜLLER, J., Regulation of Aldosterone Biosynthesis. With 19 figures. VII, 139 pages. 1971. DM 36,—

Vol. 6: FEDERLIN, K., Immunopathology of Insulin. Clinical and Experimental Studies. With 53 figures. XIV, 185 pages. 1971. DM 49,60

Vol. 7: HORTON, E. W., Prostaglandins. With 97 figures. X, 197 pages. 1972. DM 58,—

In Preparation:

BAULIEU, E. E., Bicêtre: Current Problems in the Metabolism of Steroid Hormones.

BERSON, S. A./YALOW, R. S., New York: Immunoassay of Peptide Hormones.

BREUER, H./RAO, G. S., Bonn: Metabolism of Estrogens.

CALDEYRO-BARCIA, R., Montevideo: Oxytocin.

CONARD, V., Bruxelles: Glucose Utilization.

COPP, D. H., Vancouver: Calcitonin.

EDELMAN, I. S., San Francisco: Aldosterone.

GREGORY, R. A., Liverpool/GROSSMAN, M. I., Los Angeles: The Gastrins.

GURPIDE, E., Minneapolis: Tracer Methods in Hormone Research.

JENSEN, E./DESOMBRE, E., Chicago: Receptors and Action of Estradiol.

LUNENFELD, B., Tel Hashomer: Gonadotropins.

MCKENZIE, J. M., Montreal: The Pathogenesis of Graves' Disease.

NEUMANN, F./STEINBECK, H./ELGER, W., Berlin: Hormone Antagonists in Differentiation Processes.

REITER, R., San Antonio: The Pineal Gland.

ROBERTS, S., Los Angeles: Subcellular Mechanisms in the Regulation of Corticosteroidgenesis.

STAUFFACHER, W./RENOLD, A., Genève: Pathophysiology of Diabetes-mellitus.

STOCKELL HARTREE, A., Cambridge: Purification and Properties of Anterior Pituitary Hormones.

UNGER, R., Dallas: Glucagon.